I0147679

The Elegance of Ferns

Solvejg Nitzke

Translated by Helge Dascher

the elegance of Ferns

Portrait of a Botanical Marvel

GREYSTONE BOOKS
Vancouver/Berkeley/London

First published in English by Greystone Books in 2026
First published in German in the series Naturkunden,
edited by Judith Schalansky for Matthes & Seitz Berlin

26 27 28 29 30 5 4 3 2 1

Greystone Books Ltd.
greystonebooks.com

Cataloguing data available from Library and Archives Canada
ISBN 978-1-77840-338-5 (cloth)
ISBN 978-1-77840-339-2 (epub)

Copy editing by Crissy Boylan
Text design by Fiona Siu
Cover design by Jessica Sullivan | DSGN Dept.
Cover illustration by Anne Pratt,
"Rigid Three-Branched Polypody, *Polypodium calcareum*"

Printed and bound in China on FSC® certified paper at Shenzhen Reliance Printing. The FSC® label means that materials used for the product have been responsibly sourced.

Greystone Books thanks the Canada Council for the Arts, the British Columbia Arts Council, the Province of British Columbia through the Book Publishing Tax Credit, and the Government of Canada for supporting our publishing activities.

EU Safety Information: Easy Access System Europe, Mustamäe tee 50, 10621 Tallinn, Estonia, gpsr.requests@easproject.com.

Greystone Books gratefully acknowledges the xʷməθkʷəy̓əm (Musqueam), Sḵwx̱wú7mesh (Squamish), and səlilwətaɬ (Tsleil-Waututh) peoples on whose land our Vancouver head office is located.

Contents

Introduction

Filler Greenery and a Bounty of Forms

A LOVE OF FERNS often goes hand in hand with a love of form. Having no flowers, ferns cannot rely on scent or color to seduce—nor do they need to. As one of the oldest plant groups on Earth, they have thrived for millions of years without the help of humans or animals. And yet people are still captivated by the remarkable diversity of their forms. Once you notice their intricate, lacelike symmetry, it becomes difficult to look away.

The way in which the sword-shaped fronds—the leaves of ferns—are divided depends on the species. But whether open or rolled into fiddleheads (the curled tops of young ferns), they follow the principles of fractal geometry, repeating the same patterns over and over again but at different scales. The same principle makes nautilus shells, snail shells, and pine cones so pleasing to the eye.

In *Oaxaca Journal*, his account of a fern tour in southern Mexico, the neurologist and writer Oliver Sacks notes a travel companion questioning whether one can even appreciate these plants without a knowledge of the Fibonacci sequence. Another member of the group, a fern-loving mathematician, describes them as "elegant... perfectly organized... symmetrical... complete"—even "divinely economical"—in their "realization of the simplest mathematical laws." Her portrait echoes the observations of many others throughout the rich history of fern appreciation in art, literature, landscaping, and botany. Ferns—those flowerless green forest dwellers—are endlessly fascinating, from any perspective. Their appeal is not only a matter of mathematical aesthetics. Ferns are true giants in the history of the plant world and wonders of botanical design.

Oliver Sacks sets out on the excursion to Oaxaca as a member of the New York chapter of the American Fern Society. In the company of other fern aficionados and lacking specialized botanical knowledge himself, he finds himself on "a wonderful fern adventure, with novelties and surprises, great beauty at every point." The state of Oaxaca is a place of pilgrimage for plant lovers of all kinds. One attraction is the Tule Tree, a gigantic bald cypress believed to be the stoutest tree in the world and which the

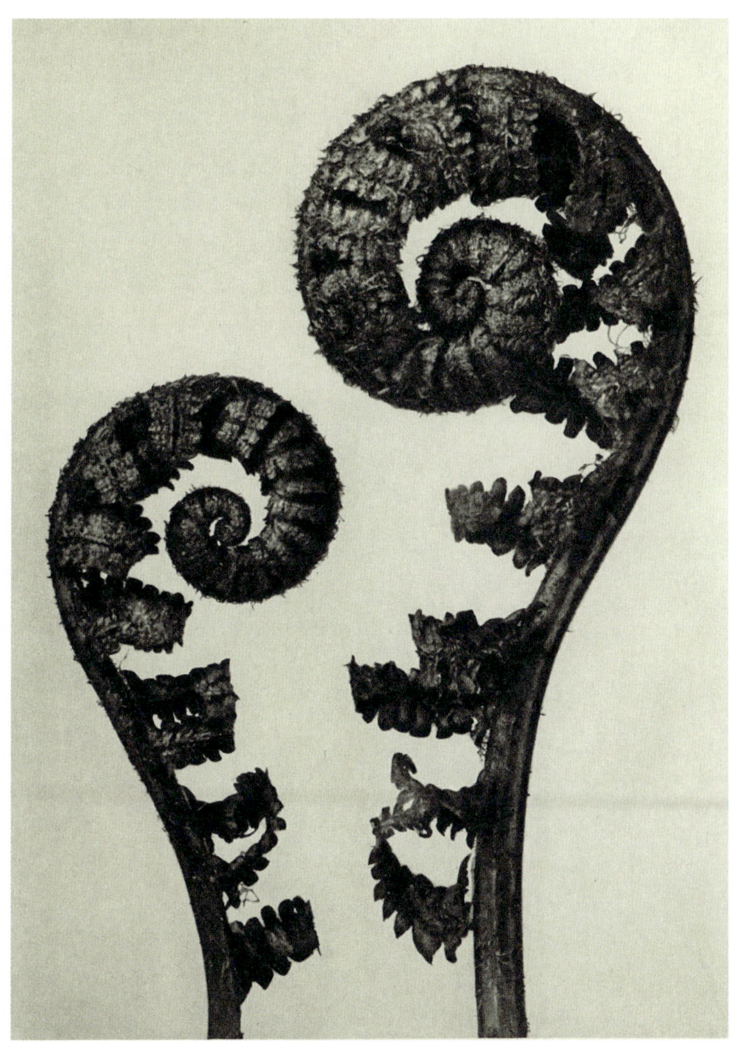

Karl Blossfeldt's photograph conveys the kinetic energy of these emerging male fern fronds. Captured just before they unfurl, they seem on the verge of exploding.

nineteenth-century naturalist Alexander von Humboldt went out of his way to visit. But it is the region's remarkable diversity of ferns, thriving in its various climate zones, that drew Sacks and his fellow travelers. In jungles and mountains, deserts and grasslands, Sacks finds ferns of every shape, size, and growth pattern—and also a sense of community with other fern enthusiasts. Despite having grown up in a fern-loving family in London whose enthusiasm spanned generations, he continues to be surprised by "how deep and passionate the love of ferns can be." As a child, visiting London's Natural History Museum, he had marveled at the evolutionary age of ferns. But the direct experience of ferns—in the company of kindred spirits, crouching on roadside embankments, examining tree trunks, searching through field guides to parse the slightest variations in leaflets—changes something in Sacks, as it might in anyone who loves ferns.

It all starts with noticing these often-overlooked plants. From European gardens to ikebana, the Japanese art of flower arranging, they are used to create a natural appearance, masking signs of human intervention. Ferns excel in the role of filler. In landscaped settings, they ease the transition between flowers and trees, provide a backdrop for spectacular blooms, and bring greenery to even the darkest places. Once the contemporaries of dinosaurs, these seemingly simple plants add a touch of the primeval wherever they grow. Almost always, they play a supporting role, tucked in beside or behind the more celebrated stars of gardens, parks, and bouquets. Even in natural history museums, ferns typically appear as dinosaur food, rarely as the main attraction.

And yet every spring, in temperate latitudes, a keen eye can observe a truly spectacular display.

It's late April, and in the otherwise unremarkable back courtyard of my apartment building, the time has come. A group of ostrich ferns (*Matteuccia struthiopteris*), planted in a dark corner mostly for practical reasons, are beginning to stir. The rolled-up fronds are still encased in a thin brown skin, which will gradually loosen as the fiddleheads, arranged in circles, push up out of the soil. The glossy acid-green shoots remind me of tentacles emerging from the abyss. Now that I've come to appreciate how remarkable ferns are, I can't help but see them as more than just green filler. Since I'm not especially good with names and, like Sacks, would rather leave the task of identification to others, I tend to focus on the more unusual aspects of ferns. These include their deep-time history, their interactions with humans in myth and literature, and especially the unusual movement inherent in their fronds. This is especially apparent in spring, when the young fiddleheads are visibly in a state of tension. They open as though enjoying a good stretch, and as their main stem—called a rachis in ferns—unfurls, you can see that each individual leaflet, too, starts out tightly curled. The fronds unroll in cascades, the fractals flowing logically out of the movement.

Last year's black-brown fronds offer a backdrop to the almost uncanny vitality of this uncoiling. Writer and poet Robert Macfarlane captures the movement in words that mirror its energy: the ferns "reach, roll and unfold" until one "*flares* . . . fully fanned."

By mid-May at the latest, the corner of the yard is a display of green fireworks. The ferns grow in tightly packed clumps, reaching skyward. They only begin to droop in summer, when their fertile spore-carrying fronds start to grow. This marks the beginning of a very different spectacle—one that is almost imperceptible, making it all the more intriguing. The way ferns reproduce was long a mystery, and even today, it is difficult to discern what really goes on beneath the fronds. While it is now standard knowledge in biology that plants alternate between sexual and asexual phases, an understanding of the stages of fern reproduction—the sexual processes that unfold in a viscous fluid under the ferns, and their relationship to romantic myths about human desire—is more niche.

In European folk tradition, ferns were considered intermediaries between the human and magical worlds, capable of bestowing extraordinary powers under very specific conditions. The *Handwörterbuch des deutschen Aberglaubens* (Dictionary of German superstitions), published in the 1920s but still a standard reference, traces the earliest mentions of German fern folklore to the twelfth-century writings of Hildegard of Bingen. The mystic and scholar recommended ferns as protective plants: houses surrounded by ferns, for example, were less likely to attract lightning, and carrying ferns offered protection against magical spells. To ward off the devil, she advised placing ferns around women in labor and infants in their cradles. Although Hildegard did not mention them, fern "seeds" were considered particularly potent. Sixteenth- and seventeenth-century herbals and grimoires described elaborate rituals for extracting the seeds from fronds on Midsummer Eve (or St. John's Eve). One supposedly effective method involved summoning the Devil by drawing chalk circles at a crossroads and placing chicory stalks in their centers. If the ritual was performed correctly, the fern's seeds would drip like resin from its leaflets.

But fern seeds could also be gathered in winter. Folklore instructed hopeful collectors to abandon prayer in favor of "devilish thoughts" during Advent, then stand at a crossing on Christmas Eve where "corpses are carried to the churchyard." The dead would rise from their graves and speak, but the collector must keep silent or risk falling under the Devil's power. Eventually a hunter would appear and shake out a bag of fern seeds, which had to be caught with a special cloth. Sources disagreed about the type of cloth, but all emphasized that the collector must not touch the seeds with bare hands.

Although the male fern is less common in North America than in Europe, its appearance is so iconic that it is practically the archetypal fern. Its preference for shade and its long-enigmatic reproductive cycle have contributed to its mystique.

The "magical properties" of ferns were well worth the effort. Until the eighteenth century, especially in northern and eastern Germany, people believed that carrying fern seeds would make them invisible. In the south and west of the country, the seeds were believed capable of attracting wealth and good fortune: they could reveal treasure or, when held in the hand, turn into gold coins. Having fern seeds in your pocket guaranteed luck in gambling, while mixing them into gunpowder ensured a clean shot. Sewing them into clothing made it resistant to cuts and stabs. If fern seeds fell into your shoes on Midsummer Eve, you could understand the language of animals. A girl who scattered them in her garden would be granted visions of her future husband. In France, a young woman could use fern seeds and flowers as love charms to capture the heart of her beloved.

In Southeast Asia and across the Pacific Island nations, ferns continue to hold mythological, practical, and cultural value, particularly in the self-representation of many Indigenous Peoples. Ferns' ability to rapidly regenerate landscapes in the aftermath of natural disasters, such as volcanic eruptions, makes them an important symbol of renewal. In many regions, particularly in New Zealand, there is a movement to revive the use of ferns as food, building materials, medicines, and decorative elements, reclaiming these practices from the margins of colonial systems and cultural appropriation.

In Germany, ferns are considered not only ornamental and mysterious but also useful. Although their medicinal uses have largely been forgotten, ferns remain valuable for erosion control, as livestock bedding, and for producing natural dyes. Here, as elsewhere in the world, the shoots (or fiddleheads) of certain species, particularly ostrich fern, are consumed as food. Careful identification is required, and they must be consumed fresh. As they may contain carcinogenic compounds, overindulgence should be avoided, but they are safe when properly cleaned and cooked and enjoyed in moderation.

Fiddleheads can only be harvested during a narrow window of time: after they have fully emerged, but before they begin to unfurl. When cut—ideally with a sharp knife—they release a distinctive aroma reminiscent of grass but richer and with a sharp, nutty quality. Preparation requires thoroughly rinsing the fiddleheads to remove their brown papery scales. Despite their tender appearance, the young shoots need to be cooked for a surprisingly long time. Some recipes recommend boiling them in salted water for up to forty minutes, but I've found that fifteen minutes is entirely sufficient. The fiddleheads can then be sautéed in butter with garlic and mushrooms or simmered in white wine. But I prefer them plain. Their green color, which only brightens with cooking, seems to intensify their flavor. Notes of asparagus and grass come through with every bite, conjuring up a sense of wild spring. I feel like I'm tasting much more than just a rare vegetable, just as I see more than filler foliage when I look at ferns.

Eating ferns offers both culinary pleasure and a connection to ancient practices. Ultimately, what we gain from these plants depends on the qualities we recognize in them. Although their magical associations have faded and their potential has long been neglected, ferns have maintained a subtle but persistent presence in natural and cultural history that hints at their far-reaching influence. Ferns present an extraordinary range of forms, nearly forgotten practical applications, and connections to Earth's creatures that long predate humans. It is worth approaching them with all our senses engaged and giving their stories the close attention they deserve.

In the Forgotten Garden

ON THE EDGE of the Dartmoor National Park in Devon, England, there's a "secret garden" teeming with fern treasures. Unlike the titular one in Frances Hodgson Burnett's children's book, this garden isn't hidden behind a wall, though it was concealed by a dense thicket for many years and only rediscovered by chance. It tells a story of forgetting in which humans are involved but in which ferns and other plants also play their parts. Only a little light filters through the high cliffs and tangled trees. Still, the ferns that grow here—that went unnoticed for almost a hundred years—are perfectly adapted to such conditions. Even under

Before Canonteign Estate was made into a landscape park, ferns were part of an almost impassable thicket covering the property. The rugged grounds inspired this picturesque view by painter John White Abbott.

the thick shrubbery, they held their own. In Burnett's *The Secret Garden*, a garden is kept hidden because it is too painful a reminder of its late owner. Yet these ferns guard their own secret: they need no gardener to thrive.

The forgotten garden lies within the Canonteign Estate, which was overgrown and almost impenetrable when it was acquired by the Pellew family in 1812. A watercolor by landscape artist John White Abbott, painted around the time of the purchase, shows fissured outcrops and gnarled trees, with only a suggestion of ferns on the rock ledges. Although neglected, the estate proved a suitable country seat for Admiral Edward Pellew, whose service to the Crown earned him the titles of Baron of Canonteign and Lord Exmouth but who never made the property his home. Its transformation into a landscape park and garden began only when the second Lord Exmouth and his wife took residence there. The project was funded by productive silver and tin mines on the estate, which were exploited until the 1880s. As they neared depletion, the third Lady Exmouth, Susan, undertook an ambitious landscaping project to protect the livelihoods of at least some of the laborers, employing them in the construction of a 220-foot waterfall. A natural cascade, previously diverted to power the mine's operations, was now redirected through a complex system of pipes and pumps to flow over an artificial rock formation, which still today offers stunning views of Devon and Dartmoor.

Did the outcrop allow for the building of the fern garden under the waterfall, or was the fernery established earlier? No one knows anymore. It was only after the waterfall became a popular attraction that the fern garden appeared in the accounts of enthusiastic visitors. In the 1880s, fern fever was at its peak, and the secret fern garden—with its rocky steps and winding paths—not only appealed to the fashionable tastes of the time but likely set new standards.

Over the course of the twentieth century, the park fell into disrepair due to a lack of funds to maintain it and, perhaps, a waning interest in such gardens. It was not until the 1990s, when the Pellews sold the estate, that the park, though still privately owned, was opened to the public to help finance its restoration.

The fern garden, however, only resurfaced in 2009. A heavy snowfall—a rare event in Devon's mild maritime climate—caused a laurel thicket to collapse, revealing the fernery that had grown undisturbed beneath it for nearly a century. The story seems almost too good to be true, but given our renewed fascination today with plants, and ferns in particular, it deserves a closer look.

Amid today's ecological crises, the fern garden as Sleeping Beauty speaks to the resilience of these ancient plants in the absence of human interference. It also raises questions about humans as stewards and destroyers of plant worlds. At Canonteign Falls, visitors encounter plants that have defied human intervention with stubborn tenacity.

ON THE BUS from Exeter that will drop me off near the park, I catch my first glimpse of Devon's lush fern landscape. As the rain pours down and the bus speeds along the narrow roads of the southwest English countryside, the hedges, which almost meet overhead to form green tunnels, blur together. It is only during the brief stops that I can really appreciate the amazing variety of plants. Bracken ferns stand taller than the hawthorn, ash,

Queen Anne's lace, and brambles, and here and there the undivided fronds of hart's-tongue fern (*Asplenium scolopendrium*) push up out of the dense greenery. In Germany, hart's-tongue grows mainly in the southwest, where the climate is mild and humid enough to support its growth, and the chalky soils and old stone walls provide the right combination of nutrients. In Devon, the perennial species is found almost everywhere, an indicator of the county's valuable woodlands and the ecosystemic importance of its hedgerows.

Here in the countryside, there are few bus stops and even fewer active routes. The connecting bus to the Teign House Inn, I discover, does not run on a regular schedule. Instead, a minibus screeches to a stop and the driver asks if I am lost. The community shuttle provides transportation for people who cannot (or no longer) drive. Before I realize it's not the public bus, I'm wedged among a group of older women on their way to the supermarket for their weekly shopping trip. Once we've established—my German accent doesn't help—that I've come to the Teign Valley not to see my *friends* but the *ferns,* they express polite surprise at the distance some people travel for such a purpose and confirm that I've come to the right place.

Thanks to the friendly lift, I arrive at the park ahead of schedule and set off alone along a pretty path around a pretty pond. In a pavilion on the shore, next to a giant royal fern, members of the local model boat club are

Common in the cracks and crevices of cliffs and stone walls, the hart's-tongue fern produces glossy undivided fronds that spill out like bright green bouquets.

getting ready to spend the day sailing their miniature vessels across the water. It's all very picturesque in a distinctly English way, but not quite my genre. I'm here because of the Victorian fern craze, not Rosamunde Pilcher, whose romance novels are more famous in Germany than in her native Britain thanks to Sunday-night TV adaptations. In Canonteign, however, the two are practically inseparable.

Just a few steps further, and the effect of the trees is remarkable. The air and the light are so different from the sunny, bright openness outside the forest that you feel you've crossed a threshold. All sounds carry the forest with them, and your own footsteps feel softer and more cautious. In Canonteign Falls, there is no wilderness as such, although the forest has been declared an "ancient woodland." Nonetheless, the well-maintained and meticulously labeled world of trees preserves a memory of what this area might have been like once, when access to the ferns was impeded not by bus schedules but by undisturbed growth.

The forest floor must have retained something of its former wildness, because here, too, a hart's-tongue fern has pushed up out of the ground: five fronds arranged in a circle. The undivided deep-green leaves gleam even though hardly any light reaches the forest floor, and they shimmer with moisture—perhaps from the last rain shower. The glossy surface is broken by slanted parallel lines that point upward from the middle of the leaf. Turn the frond over and you can see that the buckling is due to the sporangia on the underside of the leaf, which will eventually release spores to ensure a new growth of hart's-tongue in the year to come. Right now, the spore cases are still narrow and pale green, but when they're ready in a few weeks, they will look like plump brown caterpillars clinging to the back of the fronds.

The hart's-tongues are far from the only ferns growing in the greenish light of the large beech and oak trees. The wood ferns—both male and buckler ferns—are especially striking with their large, strongly divided fronds, while the perennial hard ferns arch in enormous tufts over tree roots and rocky outcrops like soft combs. Everywhere crosiers, or fiddleheads, poke out of the green, seemingly eager to unfurl into new fronds. And this opulence is not even part of the fern garden itself. Whether planted, tended, or simply tolerated, this wealth of ferns, together with foxglove and navelwort, is spread out, supreme, at the foot of the artificial waterfall, forming a delightful contrast to the formal beds of perennial and annual flowers in the lower section of the park. They seem to appeal to different senses than their tamed relatives. Perhaps because they grow in the shade, and humans have learned to distrust anything that doesn't come out into the light.

The park's layout keeps you from dwelling too long on such thoughts. You're constantly plunging into the dark

Common polypody can be identified by the small round sori on the underside of its leaves, as well as a tendency to colonize trees, especially mossy oaks.

of the forest and reemerging into clearings and lookouts—literal bright spots along the path. Finally, ninety rough-hewn steps—the so-called Victorian steps—lead to the fern garden. The light soon becomes too dim for flowering plants, and oaks line the trail instead. They don't look gnarled so much as whimsical. Their twisting branches are thickly festooned with mosses. From the path there are views of a purely green world, clinging to the steep slopes. These small windows onto fragments of the Teign Valley are a precious reminder that rainforests exist beyond the tropics. The rainforests of the temperate zones—on the Pacific coasts of North and South America, and in New Zealand, Australia, Japan, Ireland, and Great Britain—are under severe threat. Like their tropical counterparts, they are at risk of massive plant and animal biodiversity loss. But since so few people are aware of what's at stake, their decline is going largely unnoticed. Ironically, the presence of almost untouched ancient woodland here is thanks in part to private parks like this one. While the historical "enclosure" of estates such as Canonteign turned once-communal woodlands into private property accessible only to a few landowners, it also inadvertently preserved them.

The trees, mostly oaks here in Devon, are stunted and stand surprisingly close to one another. Between them everything is mossy. This makes them economically uninteresting but ecologically—and mythologically—all the richer. It is easy to imagine princesses and knights, or fairies and elves, sleeping among their roots. A unicorn wouldn't be out of place either. In Celtic tales, fairies are born in fern rosettes.

The trees look as though they are disguised. It's an impression reinforced by the polypody ferns that grow like feathers out of the thick moss. The rhizome of the common polypody (puh-LIP-uh-dee—please say it out loud!) is well covered. Only the wiry rachis—the stem—and the singly divided deep-green pinnate leaves are visible. In German, the polypody gets its common name—Tüpfelfarn, or spotted fern—from the round bumps on the top side of the leaf. Nubby to the touch, they show where the sori, clusters of spore-bearing cases, are located under the leaf. Polypodies send out single fronds here and there that peer over the mossy oak branches as if drawing their own conclusions about the world.

In the fern garden, the light changes color once more. The green brightens, but while it is less dark here than in the rainforest, a dimness persists. Out of this verdant twilight arise at last the tree ferns of the secret garden. We don't know what the original garden looked like. There are scattered descriptions in letters from Lady Exmouth's guests, but they are not as precise as today's gardeners would like. The soft tree ferns (*Dicksonia antarctica*), it appears, were added recently, but many of the "hardy" ferns, as the local winter-tolerant ferns are called, were present in the original garden.

The tree ferns in the fernery at Canonteign Falls still stand uncolonized. In the Royal Botanic Gardens at Kew, there are specimens whose trunks have been completely taken over by epiphytic ferns (ferns that grow on other plants) and look as if they are wearing feather dresses. Compared to the mossy oaks and their Kew Gardens cousins, the Canonteign tree ferns seem almost naked, but then they are quite new to this location. Here, the effects of global warming are to their benefit, and as long as the garden stream continues to flow, conditions will remain ideal for their growth. If you were to visit in twenty years' time, you might find them at their full height. At over six feet tall, they're already impressive.

The fallen oak trunks that lie next to and across the cave-like garden are completely beferned. Deer, polypody, and hart's-tongue ferns grow as if on little balconies: as far as the eye can see, there are ferns everywhere. The garden is tucked like a pocket between the waterfall-curtained cliff and the woods through which the steps lead. Tall trees shade the hollow, and the green light grows more intense, almost blinding, when the sun comes out. Ferns! Ferns! Ferns! The plantings display an astonishing variety of native and temperate ferns. Japanese painted ferns grow alongside common polypody and

Ernst Haeckel's fern illustration documents the diversity of fern fronds while also evoking an idealized fern world that has long disappeared from the northern hemisphere.

hart's-tongue ferns. You're dazzled by varieties such as silver ferns and soft tree ferns, spleenworts and hard ferns, lady, male, and other wood ferns. In every direction, everything is unfurling and fanning open. It is mid-June—clearly the best time to visit this garden.

Overwhelmed and sweaty—that rainforest feeling!—I stop to sit on a bench whose cast-iron back and arm rests are, fittingly, shaped like fern fronds. It is probably a copy of the bench produced by the Coalbrookdale Company in the 1800s. I've given up trying to keep track of all the fern species I've seen today; there are just too many. Later I receive a list, originally compiled as part of Canonteign Falls' application to the British Pteridological Society for special recognition of its fern collection. Stephen, who sold me my ticket to the park, told me that the discovery of the fern garden some fifteen years ago sparked a renewal of fern fever at Canonteign Falls. In England, an endorsement from a botanical society still has the power to attract visitors, but what really draws gardeners is the thrill of collecting and a love of form. Here, the native *Polystichum* and *Dryopteris* species in this garden have been granted National Plant Collection status by the Plant Heritage program of the Royal Horticultural Society. The pride taken in the collections, in what is now a "secret garden" in name only, is understandable. A fairy trail at Canonteign challenges visitors to spot wire sculptures of fairies and fairy doors. A giant T. rex woven from willow branches is a playful reminder of the ancient evolutionary history of ferns. But what's truly impressive is the sheer abundance of ferns. And all the green, in virtually every shade imaginable. The ferns are not merely decoration or a backdrop: they are a complete environment in themselves.

The feeling of having entered a self-contained world stays with me on my hike back to the bus stop. For over an hour, I walk through a fern environment that isn't carefully curated like the park but still clearly needs constant maintenance. If no one came with hedge trimmers to tame the ferns for a few summers, a fern jungle would spread here, though not quite as diverse as the botanical gem I've just visited. Instead, I imagine a primeval world where anything wishing to bloom would have to fight hard to push through. No wonder flowers and bees joined forces; how else could they hold their own against this green power?

It's warm and humid, an invitation to grow that ferns are only too happy to accept. Where more light reaches the ground, bracken ferns surge to dizzying heights. Darker corners are home to low-growing ferns. Evergreen species grow alongside ferns that turn yellow and brown in winter, some close to the forest floor and others as epiphytes, as opportunity allows. It's as if I were still in the middle of the fern garden. Yet on closer inspection, the

Bracken ferns spread aggressively through their rhizomes, quickly taking over disturbed areas such as clear-cuts and road embankments.

ferns don't seem so dominant after all. Here, where no one bothers to weed, they thrive in harmony with other plants, next to foxgloves, on oaks, and under ashes.

The rediscovered garden does what the best botanical collections have always done: it sharpens the eye. And a trained eye finds ferns everywhere. In Devon, and elsewhere, too, it's worth peering into abandoned wells for the delicate stems and silken leaves of maidenhair ferns, or examining the cracks of stone walls for the tiny brown-stemmed fronds of spleenwort, which reach out like tentacles even in dry conditions. Deep in the woods, careful observers can find rosettes of hard fern, while on almost all continents, tall bracken forms green seas in the summer.

Ferns and Frenzies

On Learning to Notice and Falling in Love

FOR A LONG TIME, ferns grew in the figurative margins of the books and salons of Europe. If they were not directly useful, they were simply ignored. Despite bracken's prominence in the landscape, it was so closely tied to its practical applications—as bedding, packaging, and insulation—that it was largely overlooked beyond its economic value. This neglect was also due to aesthetic preferences, dictated for generations by the meticulous order of French formal gardens and older Elizabethan knot gardens. Control was their central theme, and ferns—too unruly to be trimmed like boxwood hedges or trained like roses—appeared at most as decorative accents in containers. Only when "natural" landscape design came into fashion in the late eighteenth century did ferns begin to attract attention. Their feathery fronds embodied an informal naturalness—not to be confused with the untamed wilderness that would appeal to North Americans a century later. Ferns were just wild enough to soften straight lines and austere geometric forms without being threatening. Their form caught the eye of William Gilpin (1724–1804), whose writings would revolutionize how people thought about beauty.

An artistically minded cleric, Gilpin traveled whenever he could to the west of England and south of Wales in search of subjects for his watercolor paintings. His travel journals, illustrated with aquatints, depicted a natural world previously dismissed as uninteresting. The public responded enthusiastically. Gilpin found in these seemingly abandoned, overgrown landscapes a distinctive appeal that continues to influence our idea of natural beauty today. A key theorizer of the concept of the picturesque, he believed that picturesque beauty was not a quality inherent in nature but rather one that could be created by arranging irregular elements in a pictorial manner.

"Arranging" landscapes according to the rules of the picturesque required close observation, sensitivity to potential beauty, and the ability to represent this beauty

Instead of collecting the plentiful bracken for use as animal bedding, this young woman has gathered rare ferns to sell to urban collectors wishing to display their fern-hunting prowess.

in ways that could be appreciated even by untrained observers. Because of their wild growth habit and remarkable variety, ferns played an important, albeit supporting, role in these compositions. The picturesque gaze introduced a new range of motifs—architectural ruins, gnarled trees, rugged cliffs—that evoked longing for a vanished past and an unspoiled world. This sensibility influenced a new approach to garden and park design in which ferns were employed to create an illusion of untamed nature. Ferns also became motifs in paintings, where they mediated between foreground, middle ground, and background. The distinction between parks and natural environments—and between looking at landscape paintings and actually walking through real landscapes—became increasingly fluid, with ferns helping to blur these boundaries. Today, ferns still flourish among the follies and artfully staged old trees in the parks created in Gilpin's time, when common lands were enclosed and transformed into private estates. Because ferns grow much faster than trees, they were well suited to concealing the artifice that went into achieving an impression of naturalness. In a sense, they served as a horticultural tool for arranging parks so that visitors could wander through them as though moving inside a painted landscape.

Gilpin's *Observations on the River Wye and Several Parts of South Wales, &c. Relative Chiefly to Picturesque Beauty*, published in 1782, inspired ever more people to visit the places whose picturesque beauty he described. Although travel was becoming easier, it remained an activity for the privileged until railway transport literally picked up speed in the 1830s and 1840s. The less affluent also journeyed through these landscapes, but their reasons for doing so and the range of their trips differed dramatically from those of wealthy tourists seeking an aesthetic education.

In 1820, more than a generation after Gilpin's *Observations*, a volume of poetry appeared that challenged picturesque conventions. It captured the imaginations of city dwellers, becoming a literary sensation. *Poems Descriptive of Rural Life and Scenery* by John Clare (1793–1864) promised a poetic journey through the countryside, an unfiltered view of human nature, and the pleasure of reading something both extraordinary and deeply ordinary. Clare, a farm laborer with no property of his own, was born in Helpston, Northamptonshire, around one hundred miles north of London. Despite his poetic ambitions, he spent little time at school, and his relationship with both grammatical and social conventions was, to put it mildly, complicated. While naturalness was celebrated in art and landscape, the prevailing norms allowed little deviation, and Clare lacked the financial means to simply ignore them. Still, poetry held him in its grip. He read whatever he could find and discovered kindred spirits in the Romantics and earlier rural poets.

Ferns excel at companion planting, naturally forming attractive communities with heather and juniper, as seen here, as well as with digitalis and many other flowering plants—all without the assistance of human gardeners.

Unlike Wordsworth and Coleridge, whose excursions in the Lake District took them past royal ferns (*Osmunda regalis*) and hart's-tongues (*Asplenium scolopendrium*) growing alongside picturesque waterfalls and rock formations, Clare lived in a flatter and more provincial environment. The wilderness that unfolded there, between small fields, enclosed orchards, and woodlands, was more modest than in northwest England, but Clare perceived it as few others could.

Clare became famous as both a peasant poet and a critic of botanical practices. He considered it absurd to cut plants and dry them to create herbariums filled with mummified specimens, labeled with Latin names and classified according to what he called Linnaeus's "dark system." Why settle for dead plants when you could venture out and discover living ones in their natural habitats? Jonathan Bate, Clare's biographer, writes with some sympathy about Clare's frustration at his teachers' attempts to educate him in grammar and basic science. Clare didn't see the point, suggests Bate, since he never doubted that he knew nature—which he could explore by walking—better than he knew himself. He did, however, doubt his place in human society; he felt more at home among wildflowers and woodland creatures. In his poems, he addressed larks, buttercups, and wild roses by their common names, treating them as friends. He saw himself and his "simple" nature reflected in them, and he seemed continually torn between embracing a literary life and keeping to his familiar world. Poetics, grammar, spelling, and taxonomy were all secondary: what mattered to Clare was the direct encounter with the plants themselves.

Unpublished during his lifetime, Clare's early poem "To the Fox Fern" describes one such encounter. In the dark stillness of the woods, the hard fern (*Blechnum spicant*), whose fronds somewhat resemble foxtails, seems to come alive. "Haunter of woods, lone wilds and solitudes," it seeks out or conjures a strange mood far removed from the bright world of fields and meadows. The poem follows the sun's light as it filters through the forest canopy, creating around the fern "a quiet dimness fit for musings / And melancholy moods." This shadowy, poetic atmosphere clings to the fern's fronds, and it seems to accompany Clare beyond the poem itself. He, too, was inclined to "musings" and "melancholy moods," which he dulled with drink or processed in episodes of compulsive writing. The contrasts that marked his life—between his literary ambitions and modest circumstances, his political convictions and the need to satisfy the expectations of his readers and patrons, his depressive and manic phases—finally became too much to bear and he "went mad," as they said at the time. He began experiencing delusions and sometimes believed he was Lord Byron. Nonetheless, encouraged by his doctor, he kept writing.

Some of Clare's most powerful poems date from his time in two asylums, one in Epping Forest, the other in Northampton, where he spent the last three decades of his life. The poems are silent about his experiences in these institutions, and Clare himself could hardly maintain a conversation. But he continued to develop as a writer, even though doctors blamed his "addiction to poetry" for his illness. In these later poems, nothing stands between Clare and nature any longer. "All Nature Has a Feeling" praises a "green life": woods, fields, and brooks "speak happiness—beyond the reach of books." And in "Sighing for Retirement," Clare again describes his response to ferns. His love of "brakes and fern" reflects a view of nature focused not on the aesthetic value of bracken and its wildly growing fronds but on their abundance and the beauty of their seasonal transformation, when "pleasant Autumn comes, / And turns them all to brown." The poet was aware of the uniqueness of his perspective and, unlike Gilpin, seems to have had no ambition to impose it on others. It was enough for him to recognize what he was looking at, where others saw only thickets:

To common eyes they only seem
A desert waste and drear;
To taste and love they always shine,
A garden through the year.

It would be easy to describe Clare's writing as a precursor to the phenomenon that would sweep first England and then the Commonwealth: fern fever. He let ferns flourish in his garden, celebrated them in his writing, and admired their wildness. But for this very reason, he would probably have rejected the collecting craze that spread throughout the country while he lived—protected, it might be added—in the asylum.

Several factors converged to spark the fern fever that gripped the British Empire until the First World War. Sarah Whittingham is the foremost historian of this period. Her definitive illustrated book *Fern Fever: The Story of Pteridomania* (2012), now unfortunately out of print, offers substantial evidence that the fern craze was not just a fad but a collective mania. Stories of theft and betrayal, tales of intrepid plant hunters, and the emergence of a whole industry of fern products—from books and vascula (cylindrical containers used to collect plants in the wild) to furniture and clothing—show that the symptoms cut across all social classes.

By the mid-1850s, the infection was so widespread that the Reverend Charles Kingsley wryly coined the term *pteridomania*, or fern mania, to diagnose a condition that, he claimed, affected mainly young women—at their fathers' expense. The aesthetic revaluation of "wild" landscapes and ferns in the picturesque and Romantic traditions was a prelude to this. But it was technological change that unleashed the craze, as railway development and the shrinking distances between industrialized cities enabled the newly emerging middle class to travel in search of beautiful views. The prospect of combining an aesthetic education with an interest in science, especially botany, made such travel an acceptable activity. Young women were encouraged to assemble herbariums and take up the fine art of drying and pressing plants. The genuine scientific interest and remarkable technical skills that often characterized their engagement in these "ladylike" activities were routinely ignored.

One woman who exemplified this fusion of botanical interest and aesthetic sensibility was Anna Atkins (1799–1871). The daughter of the scientist John George Children, she spent years illustrating his work with precise botanical and zoological drawings before turning to photography—a medium she discovered through her father's acquaintance with one of its inventors, William Henry Fox Talbot. The cyanotype process, also known as the blueprint process, was developed by the astronomer and polymath John Herschel in the early 1840s. Atkins used cyanotypes, or photograms, to create prints that depicted plants with unparalleled accuracy. In this process, plants are placed between a glass plate and paper that has been treated with an iron solution. The plants are then exposed to direct sunlight. On a clear, sunny day, it takes about half an hour for the solution to react with

Anna Atkins received fern specimens from around the world for her "blueprints." The cyanotypes she created reveal the plants' intricate forms to striking effect.

the light, forming an insoluble blue pigment that stains the exposed areas. When the backing is rinsed, a delicate white silhouette remains in place of the plant. Alongside her renowned algae prints, Atkins also produced pictures of ferns, whose fractal forms stand out in exquisite detail against the deep-blue background. Atkins recognized the scientific value of these extraordinarily precise images and donated copies of her books to scientific societies. These institutions, however, all barred her from membership because of her gender, with the result that she was almost erased from history.

John Clare and Anna Atkins shared an eye for ferns and for the wonders of the ordinary, and both nearly fell into obscurity. Clare was rediscovered in the twentieth century as a poet's poet, while Atkins is now recognized as one of photography's pioneers. Their work resonates today because it teaches us ways of seeing that matter more than ever amid our current ecological crises. Notably, both were working at the very beginning of the industrial acceleration that set these crises in motion. Its impacts eventually reached even ferns: while Clare was writing in an asylum and Atkins was perfecting her cyanotypes, a physician in the heart of London made a discovery that would finally bring fern fever to the masses.

Nathaniel Bagshaw Ward (1791–1868) was a doctor by profession and a botanist by calling. In the garden behind his home in London's East End, he tried to grow ferns on a rock wall. Like other urban gardeners, he had little success, despite redirecting a pipe to let water trickle over the wall. Soot from the city's countless chimneys not only settled on hanging laundry but also suffocated sensitive plants. Frustrated, Ward gave up and turned his attention to other experiments. One would change the course of the nineteenth century.

In 1829, Ward placed the chrysalis of a sphinx moth (*Sphingidae*) on a bed of earth and leaves inside a wide-mouthed glass bottle to see if it would develop. Not only did the moth hatch in the sealed container, but two green sprouts emerged from the soil: a stalk of meadow grass (*Poa annua*) and a male fern (*Dryopteris filix-mas*). Inadvertently, Ward had found a way to create an atmosphere in which ferns and other organisms could thrive, protected from harmful external influences. As a well-connected amateur botanist, he recognized the potential of his discovery: not only would it allow him to establish the fern garden he had always dreamed of, but if adapted, it could revolutionize the transportation of live plants. At the time, only around one in a thousand plants survived the ocean voyage from the colonies to Europe. This was often sufficient to transplant plants such as coffee, cacao, and fever trees (*Cinchona*) from one colony to another, and later to ship their products (fruits, seeds, or, in the case of the fever tree, bark) to markets around the world. But stories of adventurous plant hunters who risked their lives

Wardian cases were portable greenhouses that protected the atmosphere needed by ferns and other plants to survive long-distance transport. Victorians also used this breakthrough invention to grow exotic tropical plants in their own homes.

to steal plants from forests and gardens, carefully wrapping them in moist moss for the journey to Europe, rarely mention that most of these shipments arrived at their destinations carrying little more than the moldy remains of their green treasures.

The so-called Wardian case was truly revolutionary. Consisting of a transport crate with a glass roof and a wooden frame, it was designed to withstand the rigors of ocean travel. In his book *The Wardian Case*, the historian

Luke Keogh, who has tracked down the last surviving examples, describes these marvels of plant logistics as "portable greenhouses." In 1833, in the first test of his new invention, Ward asked a scientist friend to take a case filled with ferns and other plants on a voyage to Australia and to return it filled with native Australian plants. Months passed before Ward received confirmation that his experiment had been successful. By the time the letter from Sydney reached him, the ship was already on its return voyage. Ward took possession of his case a few weeks later. Inside was a "delicate coral fern" (*Gleichenia microphylla*) entirely unknown in England. The experiment's success marked a breakthrough of immense scientific and economic value. Keogh notes, "As a technology, the Wardian case provided the possibility to move plants; as an enclosed case it moved more than plants—it moved environments."

But Ward was interested in more than exotic ferns. It seems that he cared little about the commercial potential of his invention—he never even sought to patent it. Instead, he was captivated by the prospect of creating urban environments hospitable to ferns. Like Clare and Atkins, Ward had a particular passion for native plants, especially ferns, admitting that "if some groundsel or chickweed had sprung up in my bottle instead of the fern, it would have made no impression upon me." Following the incident of the fern in the bottle, he experimented with various forms of small glass cases. They soon filled his entire home, and when space became scarce, he built additional greenhouses in his garden.

Greenhouses had been in use in England since the eighteenth century, but it was the new materials and engineering techniques developed in the nineteenth century that enabled the construction of palm houses large enough to house the exotics that Ward's cases now brought to the British Isles. Ward contented himself with a version that was smaller and yet no less elaborate than the structures that would delight Victorian society about a decade later at the Royal Botanic Gardens at Kew. His model—eight feet square and built outside a staircase window—was not so much a hothouse as an atmospheric barrier that shielded the plants from soot and pollution. It also allowed the doctor, who witnessed the effects of this pollution on his patients daily, to enjoy a reprieve from the city air.

Ward named the greenhouse Tintern Abbey, after the famous monastery ruins in South Wales. It housed a picturesque fern world that he enjoyed showing to neighbors and friends. His backyard oasis anticipated a fashion that would become a fixture of Victorian parlors, which were as sooty as the streets. Elaborate glass cases and bells with drainage systems stood on custom-made stands as parlor centerpieces or were attached to windows, keeping middle-class women busy and providing their families with more pleasing views than those outside.

Ward believed that such tableaux offered numerous benefits, even if the healthier atmosphere inside the miniature greenhouses could not always be accessed directly. Parlor gardens provided a sense of comfort and serenity, he suggested, and the mere sight of ferns in their varied forms had such positive health effects that he recommended them for hospitals, especially sanatoriums. If he had filed for a patent, we might assume he was driven by business interests. Instead, he appears to have been genuinely motivated by the desire—as optimistic as it was sad—to provide an alternative to the vistas the urban world had to offer at that time. It's hard to imagine someone like John Clare, who associated ferns with the dreamy, melancholy mood of woodlands and wild bracken, not falling into deeper despair at the sight of enclosed ferns. For Anna Atkins, too, tending a miniature garden instead of carrying out her precise work with plants in bright sunlight would have been a kind of punishment. Yet the approach perfectly suited the Victorian desire to exert total control over the world's diversity, even in the smallest spaces.

The glass cases in Victorian parlors seem tiny compared to the gigantic undertaking that occupied Kew Gardens' architects, engineers, gardeners, and botanists. Although one of the most elaborate Wardian cases was actually built to resemble that massive project, the difference in scale forbids comparison.

On the walkways of Kew Gardens'
Palm House, visitors could experience the
thrill of the jungle without its dangers.

The historian Kate Teltscher describes the enormous Palm House at Kew Gardens—an engineering triumph, constructed between 1843 and 1848—as a masterpiece of nineteenth-century botanical architecture. In her book *Palace of Palms*, she presents it as "a glittering prism through which to view Britain's real and imagined place in the world." Despite the addition of several equally imposing greenhouses in the more than 150 years since its opening, the Palm House has retained its original grandeur and continues to fascinate visitors to this day. This is due not only to its historical significance and charm but also to the reinvention of the surrounding gardens that it inspired.

In 1840, when the recently crowned Queen Victoria decided, as a cost-saving measure, to "generously" cede the gardens to the British government, the old conservatories were overcrowded, and the beds of the then much smaller garden were in poor condition. This was a catastrophe for the world of botany. The collection that had been assembled in the late 1700s under Joseph Banks, arguably Kew's most influential director, was unparalleled. Yet by the early 1800s, its splendor was hardly evident. The palm trees had grown so vigorously that they had pierced the fragile glass roof of their enclosure, endangering the collections. Moreover, the gardens were mostly closed to visitors because the gardeners feared the presence of the public might destroy what little order remained.

Everything changed under the direction of William Jackson Hooker, beginning in 1841. The gardens were redesigned and opened to the public during designated hours. Working with Kew's curator, fern specialist John Smith (1798–1888), as well as numerous landscape designers, architects, and gardeners, Hooker developed an arboretum, large ponds, and radial paths, all centered around the new Palm House. Even from the outside, the Palm House commands attention, rising above the ponds, trees, and shrubs with its weathered white iron framework and seemingly opaque panels. The sensation of entering a completely different tropical world captivated Victorian society, and remains powerful today. With its sweltering heat and humidity, the structure is very much a "hothouse." The building's lofty height is obscured by masses of large-leafed plants, so the first impression is not one of scale but of the extraordinary diversity of the vegetation brought together in this space. For more than 170 years, Teltscher writes, it has provided "the ultimate spectacle":

> You can wander through the tropical regions of Africa, the Americas, Australasia, Asia and the Pacific in less than an afternoon. Embowered in vegetation, you can easily lose all sense of direction. The paths seem to loop endlessly, and the ever-changing views from the winding staircases disorientate you further. It is a manicured jungle: the leaves

> are swept up and the fruit is harvested, but the stone-flagged floor is puddled and branches reach out over the paths. Peering into the beds, you might spot robins or, more rarely, a Chinese water dragon that keeps the number of cockroaches down. Removed from the normal constraints of time and place, the Palm House is both inside and outside, real and staged, here and there.

The enormous structure is physically bounded, yet it creates the impression of containing entire landscapes. Maybe that is because the plant varieties are as diverse as their geographical origins: although most species are represented by only a few specimens, you feel completely immersed in a world of greenery. To emerge, you climb one of the winding staircases to an elevated walkway. The canopy spreads out below like an expansive model of the world's tropical flora. The view is awe-inspiring not only because of the tree ferns, palms, and refined architecture but also because of the elevated perspective. From above, you look down upon a deliberately arranged landscape—an emphatic display of mastery over nature. By enabling plants gathered from around the world to thrive in cool, rainy London, not merely preserving them as dried and pressed herbarium specimens, Kew's nineteenth-century gardeners created a showcase of botanical expertise and geopolitical power—a living display made possible by the glass-and-steel roof above.

The botanical marvel was matched by an engineering triumph. Calling the Palm House an achievement of local craftsmanship would hardly do justice to this monument to Britain's industrial supremacy. Just as the selection of plants testified to jungles and fern forests on the other side of the planet (provided the wild populations had not been entirely depleted), the building itself embodied an industry poised to transform everyday life. Coal-heated glass-and-steel structures would go on to become the future of construction, but they first entered the everyday lives of the emerging middle class in the form of greenhouses. The huge conservatories were ideal advertisements for the mass-produced glazed boxes that, in the second half of the 1800s, brought miniature versions of these hothouses into English parlors, onto English windowsills, and into the windows of English psychiatric institutions.

When it comes to ferns specifically, the Temperate House, located just a short distance from the Palm House, is almost more fascinating. Because temperate plants prefer cooler temperatures, this structure could be made even larger than the older showpiece, with twice the floor area. Construction began in 1860, although the building did not open for another forty years. Restored between 2011 and 2018, it now boasts a fern collection unequaled in the world. The number of ferns in the Kew collections increased more than a hundredfold between the 1820s and the 1880s, laying the foundation for the spectacular

Fern collecting was as much a social activity as a botanical one, providing one of the few socially acceptable opportunities for unmarried young women and men to spend time together.

display on view today. The fern trees from Australia and New Zealand, which resemble palm trees at first glance, are especially striking, in part because their fronds are so finely divided and because many are entirely overgrown by other ferns. During the renovations, the collection's labels were also updated to reflect today's concerns and interests. *Biodiversity* and *climate* are keywords, and part of the exhibition explores how the ferns came to be in the Kew Gardens in the first place. A history of the Wardian case quickly veers into criticism of colonial exploitation and a commitment to improved practices. Today, Kew runs several biodiversity programs worldwide aimed at protecting fern populations and restoring species once destroyed by generations of its own plant hunters.

The Victorian fern craze shows how botanical trends can cut both ways. While people were inspired to notice and appreciate plants, it is unclear whether this attention benefited or harmed wild fern populations, or what lessons we should take heed of as interest in these plants grows again today. At the time, "serious" fern enthusiasts complained that pteridomania would lead to careless tourists destroying fern populations and even wiping out native species, but they were likely not entirely sincere: their real frustration was probably with young couples seeking privacy in secluded spots where rare specimens grew.

In his book *Flora Britannica*, the botanist and writer Richard Mabey suggests that the development of the railways—so important to fern tourism—may in fact have enabled ferns to spread more easily. Steam locomotives transformed railway beds and their stonework into warm, humid habitats that were hospitable to ferns, and the trains supplied "a means for ferrying their spores about." Mabey is not a fan of botanical fads, but he observes that intensive collecting probably did not cause any fern species to become extinct. This does not mean, however, that ferns are safe today. As Kew's gardeners point out, habitat destruction is more often the result of indifference than of excessive enthusiasm.

WHILE THOSE WHO considered themselves to be "real" fern specialists would have preferred to keep the masses out of fern habitats, others worked vigorously to spread fern fever. Just as botanical gardens were intended to provide the public with "rational recreation," fern enthusiasm was also considered a form of meaningful entertainment. In the nineteenth century, the focus was on public health, not ecology and conservation. Like the physician Nathaniel Bagshaw Ward, the writer and fern enthusiast Francis George Heath (1843–1913) believed that the appreciation of ferns had positive health effects. Heath was a central figure in the fern book market for decades, supplying it with fern guides, lavishly illustrated portfolios, and handbooks. His bestseller *The Fern Paradise: A Plea for the Culture of Ferns* (1875) was an impassioned appeal for people to pay attention to ferns. If a book lacked inspirational or literary merit, Heath wrote modestly, it must at least be useful.

> The earnest purpose of this volume, then, is that it may assist in developing the popular taste for Ferns in such a way as to lead to the more extensive cultivation of these graceful and beautiful plants in our gardens and in our dwelling-houses; nay, even so far as such an arrangement would be practicable, in our places of business, wherever they may be.

The ambitious project gained urgency from problems that were becoming all too apparent. Not only were cities

The higher the print run of fern books, the more opulent the design. Francis George Heath's *The Fern Paradise* (1875) included photographs and cyanotypes, making the most of cutting-edge printing technology.

becoming more polluted, they were also growing darker and more densely populated. The urban poor had few options to brighten their bleak surroundings and satisfy their longing for nature.

Heath did not push for far-reaching social reforms. Although he complained that London lacked sufficient parks to provide residents with adequate "breathing space," he was not primarily concerned with improving working-class neighborhoods or ensuring access to sunlight and clean air. As a quintessential Victorian, he instead reminded readers "how much in this life happiness and misery, comfort and discomfort, depend upon ourselves and upon acts or habits that are within our control." Rather than complaining about darkness and shade, he recommended finding pleasure in ferns, which could brighten "every room in a house" and "every vacant and available corner." According to Heath, the main obstacle to this obvious solution was that the existing fern literature was too difficult for ordinary readers to understand.

In his attempts to popularize fern culture, Heath did not hesitate to challenge the experts. One target of his contempt was the naming mania that accompanied fern fever. M. C. Cooke's *Fern Book for Everybody* (1867) listed eighty-five variants of hart's-tongue fern alone. A catalog compiled by the head of the British Pteridological Society in response to a request for simplification listed more than a thousand. Heath found this unacceptable.

The "piling on" of Latin names irked him as well. He preferred to use descriptive English names of his own creation, replacing *Scolopendrium vulgare crispum-latum*, for instance, with "broad curly hart's-tongue."

In *The Fern Paradise*, Heath begins by inviting the reader on a "ramble" before delving into fern species and cultivation techniques. He does not, as the frontispiece suggests, take the reader to remote, exotic locations but to Devon—in his words, an unrivaled "fernland." His descriptions are captivating and richly informative. Surprisingly, today's rambler can still find "green lanes" that look very much as Heath described them.

> Onward, still onward, and downward winds our lane, until, all at once, it becomes fairly buried under the glorious mass of vegetation which grows with such wild and beautiful luxuriance around and above. We have now almost to crawl underneath the bushes and the graceful Fern-fronds which literally choke up the way. [. . .] Grand as we have hitherto found the development of Fern-life, here, in this spot, we find the grandest development of all. Oh! The keen enjoyment we derive from the delicious coolness of this almost subterranean avenue!

He rattles off the names, shapes, and growth habits of ferns with such enthusiasm that you feel you're right there with him. Heath acknowledges that fern hunting, like any new hobby, can initially seem overwhelming, but he assures beginners that they will quickly find their bearings. After reading the book—especially if you've ever walked in Devon on a warm, humid midsummer's day and experienced the abundance of ferns—you'll likely find yourself agreeing with Heath: perhaps people really should grow a few ferns to improve their physical and mental health.

From edition to edition, Heath's instructions and explanations for fern cultivation became increasingly elaborate. While his book was initially aimed at bringing a bit of joy to those who had almost no access to nature, by the fifth edition (1878), his recommendations had evolved to include considerations of the "hygienic influence" of ferns on rooms and their contribution to air quality. Heath recommended glass boxes on a porcelain base, which could be opened so that even "in midwinter" one could breathe the sweet scent of fresh greenery. He raved about window gardening, which allowed ferns and ivy to bring life to even the dreariest homes. The most ambitious enthusiasts, he suggested, could build fern rockeries and caverns. A small fern grotto, for instance, required only the installation of a pipe in the yard, with a small waterfall to keep everything moist.

> The communicating pipe should be brought to the centre of the "well" floor, and provided with a tap and spreader.

> Around its base a cluster of small rocks can be cemented. Upon the sides of the "well" and at the top more rockwork may be constructed, so as to form a cave closed on every side, save the one facing the room. At the top a small aperture must be left, in order to admit some light, and to give ventilation.

Heath recommended planting ferns all over the rockery as a visual complement to the sound of the flowing water. Conveniently, the back of his books included advertisements for fern nurseries and rockery builders. With "no reason, indeed, why [such caverns] should not be constructed . . . in any rooms of the house; and even in mid-room," demand was practically guaranteed.

Heath's vision for bringing ferns into even the darkest London apartments—brightening the rooms of the "poor seamstress in yon ill-lighted garret," whose flowers only wither "where the glorious sun never comes"—drew skepticism from reviewers. After all, people struggling to put food on the table had no time or money for such frivolity, and ferns were only truly beautiful in nature. A stunted fern in a squalid part of London in poor weather was more likely to inspire melancholy, wrote one critic. Heath, ever pragmatic, remained unperturbed. He couldn't imagine a fern inspiring any feeling other than delight, regardless of location. If "looking at Ferns alone in gloomy weather" prompted such a response, the person affected should "by all means" look away, Heath wrote. He promised, though, that anyone who observed, cultivated, and cared for ferns would experience an improvement of mood and outlook. He admitted that his book was aimed more at philanthropists than their beneficiaries, but he made his appeal clear: those who could rarely get out to the countryside were the ones most in need of ferns.

Heath wrote during a time of rapid industrialization and growing fears of social unrest. While some people strolled through parks in fern-patterned lace shawls or sipped tea from fern-decorated cups as they planned weekend expeditions and compared botanical finds, others worked in factories six days a week from childhood. Heath was not blind to this inequality. He recognized that the alienation from nature that he observed among the "humbler classes" was not due to lack of interest or ability. Instead of blaming or ignoring working men and women, he sought ways to bring nature into urban life.

The problem has only deepened with time, however, as poverty has persisted and the gap between everyday urban life and the natural world has widened. The sight of plants is still held up as an antidote to the harmful effects of modern life, offering respite between occasional escapes from the city. But thinking back to John Clare, it seems doubtful that ferns alone can heal the rifts created by the modern world. The longing to immerse ourselves in "brakes and fern"—to find poetry in fields—points to a greater utopian freedom than a glass-box garden or parlor grotto can provide. And yet perhaps the attempt is worth making anyway.

Ferns challenge the identification skills of even advanced botanists. Like hidden picture puzzles, they yield their secrets only to patient observation.

Identifying and Naming

A Short Fern Primer

HOW CAN FERNS be distinguished from other plants and from one another? The question is hardly trivial. The class of true ferns (Polypodiopsida) encompasses some 10,600 species of ferns, horsetails, and adder's-tongues, found in almost every ecosystem on Earth. These were formerly grouped with the related but geologically older club mosses (Lycopodiopsida) as vascular spore-bearing plants because they both reproduce via spores, like regular mosses, and transport water through vascular tissues, like seed plants. But it turns out that they are not as similar as these shared traits suggest. Ferns, with their more complex structure and greater diversity—there are only about 1,300 extant species of club mosses, compared to the over 10,000 fern species worldwide—are in fact more closely related to seed plants. The "fern seeds" and "fern flowers" of myth and folklore suggest that this relationship was intuited long before it was scientifically established through plant systematics.

True ferns grow on land (terrestrial) and in water (aquatic), on stone (lithophytic) and on other plants such as trees and tree ferns (epiphytic). They can be perennial or annual, herbaceous or treelike, and some have a form that resembles four-leaf clover. Their adaptability means that ferns can grow almost anywhere on the planet, and the fossil record shows that they once grew in places like Antarctica that are now virtually devoid of plant life. But since the spread of conifers and flowering plants across the globe, ferns have rarely dominated the landscape. To study ferns, therefore, we must first learn to recognize them and distinguish them from other plants.

In his books, Francis George Heath, the great fern popularizer, intentionally delayed addressing the question "What is a fern?" until several chapters in. Describing his rambles along Devon's country lanes, he suggested that knowledge of ferns was best obtained not from reading botanical and other scientific books but by direct observation, which by itself could lead to an intimate understanding of these plants.

Nomenclature was secondary for Heath. He preferred to call ferns by their common names (hart's-tongue, fox fern, and polypody), mentioning their scientific genus names (*Asplenium, Blechnum, Polypodium*) only once, if at all. Heath regarded his contemporaries' tendency toward excessive classification and elaborate naming with dismay. He opposed coining new names for ferns that showed only slight variations and deplored "absurdly long and unpronounceable names," an "affliction" also criticized in an 1890 issue of *The Garden* magazine. This naming frenzy was partly due to the ease with which ferns adapt and change, resulting in a wide range of frond shapes within a given species as well as in numerous and sometimes very different mutations from which new varieties can be bred. As the "discoverers" competed to produce new and exciting varieties, they outdid each other in coming up with complex names, such as *Polystichum angulare* var. *divisilobum plumosum densum*. Heath wondered who could remember such names and whether they contributed to meaningful conversation. More importantly, how could someone struggling with such terminology enjoy all the benefits of getting to know ferns? Heath left the complex names to the botanists; he believed ordinary fern enthusiasts needed a simpler way to explore and enjoy the "fern paradise."

This was not a rejection of botany or taxonomy but a deliberate stance: there are things worth knowing about ferns, and things that, though knowable, risk distancing people from plants. Heath focused on fern species that were well suited to cultivation or commonly found in England. When, in *The Fern Paradise*, he finally gets around to defining ferns, he charmingly acknowledges that he is going to borrow a "hard word from botany" ("just one"!), unearthing from the depths of the taxonomic system the term *Cryptogamia*. The fern, he explained, "is a flowerless plant. Although flowerless, it is not seedless; but its peculiarity is that it acquires its seeds without the intervention of flowers." Much could be said about fern seeds, but Heath left it at that. What mattered to him was that as members of the class Cryptogamia, ferns cannot be identified by flowers. Concerned with his readers' moral and physical betterment, he quickly moved past this slippery patch of fern science—no additional information was needed for the level of identification he intended. He simply wanted people to be able to recognize ferns, an ability that he was convinced should come easily.

And it's true: with a little practice, anyone can learn to tell the difference between fern fronds and the stems and leaves of flowering plants. You don't need to know all ten-thousand-odd fern species—or all the native species and hybrids of a single area—to recognize ferns by their distinctive features: the typically elongated shape of their fronds, their arrangement, their preferred habitats. If you see a flower or, in the case of the misleadingly named

All parts of a fern, from the rhizomes and sporangia to the smallest leaflet, have distinctive characteristics that help in classification.

palm ferns (order Cycadales), a seed-bearing cone, you're not looking at a true fern. But if you see small dark spots (sori) on the underside of fronds, you know that you've found one.

The workshops run by the British conservation organization Plantlife aim for about the same level of familiarity. In March, just before the start of the new growing season, it holds a "Spring Into Action" week. Plantlife is not a botanical society; rather, it was founded to raise awareness of the value of wild native plants and the importance of protecting them. Plant identification is an essential part of its program. The identification workshops focus on basic orientation rather than expert knowledge, helping participants spot often-overlooked plants. Through this kind of attention training, Plantlife wants to motivate members and other plant enthusiasts to start noticing differences.

Like Heath's guidebooks, the workshops provide just enough fern-specific vocabulary to help people identify the plants they find. This approach is central to the organization's mission as a plant advocacy group. Most of us learn to identify animals from childhood—we can tell squirrels from mice, sparrows from robins—but our botanical knowledge often stops at the most common trees. What we don't notice, the organization believes, we can't protect. One hundred and fifty years after Heath's popular books on ferns, Plantlife is once again issuing a passionate call to action. Today, however, the challenge is not to win over a society uninterested in botany or unaware that potted plants can thrive on a windowsill but one completely disconnected from wild plants.

I attend an online workshop led by Rachel Jones, whose work at Plantlife is dedicated to increasing the resilience of woodlands. She begins by explaining how to spot a fern. Ferns, as we've learned, are plants without flowers; instead, they have capsules containing spores. In Europe, they tend to be low-growing. This doesn't mean you shouldn't look up, she says, holding a polypody frond toward the camera. After all, some ferns like high places. A photograph from a morning walk last autumn shows an oak branch covered in thick moss. Polypody ferns poke out of the moss, rising from the tree into the damp, misty air. It's like a look-and-find puzzle: the longer you study the photograph, the more ferns and fern species you spot. Rachel's enthusiasm is contagious, even in this online format. Although it's March and there aren't many ferns to be found this early in the season, you can sense how much she'd like to be outside, pointing out rhizomes and last year's fronds.

To equip us with the skills that will transform our next outings into journeys of fern discovery, Rachel introduces us to the basics of fern identification. She distinguishes between frond shape and growth habit. Rosette-forming ferns such as lady ferns (*Athyrium filix-femina*) grow in compact funnel-shaped clusters. Creeping ferns such as bracken (*Pteridium aquilinum*) spread through long rhizomes, producing individual fronds that can attain impressive heights. Their running rhizomes give these ferns an especially broad reach, allowing bracken to colonize large areas with incredible speed; such invasions can be seen where spruce stands have been clear-cut. Ostrich ferns (*Matteuccia struthiopteris*) combine the two growth habits: their funnel-shaped crowns send out rhizomes, allowing them to spread even when spore dispersal fails. Better safe than sorry.

Frond shape allows more precise identification of the species, but this is where things become significantly more complex. From the rhizome, or rootstock, which lies mostly underground, grow the petioles, or stalks, that carry the fronds—ferns' leaves. The leaf blade—the leafy part of the frond—can be undivided (like hart's-tongue), pinnate (singly divided, like the hard fern), bipinnate (twice-divided, like the male fern), or tripinnate (thrice-divided, like some lady fern varieties). Some ferns are even divided four or more times. Botanists further distinguish between pinnate fronds (with leaflets divided to the midrib) and pinnatifid fronds (lobed but not fully divided). This leads to compound descriptions like "pinnate-pinnatifid frond." The possibilities don't end there.

In addition to growth habit and frond shape, a fern's sporangia (spore cases) and their arrangement in sori (clusters of spore cases) can also help with identification. Are the sori located on the underside of the fronds or, in the case of fern species with sterile fronds, on sporophylls (fertile and often non-photosynthetic fronds)? Are they round, linear, or kidney-shaped? Are they packed tightly together or distributed across the frond's lower surface? Do they form raised bumps, as in the case of polypody ferns? Even the indusium, a protective membrane that covers the sori in early summer, differs by species, as do the scales and hairs on the rachis (the stem or midrib of the blade). Habitat matters too. Where does the fern grow—on the ground, on a stone wall, on a tree? How big, how close to one another, and how green are the fronds? Are they winter-hardy or not? Is the location sunny or shaded? Is the soil acidic and moist or calcareous and dry?

Some answers are so surprising that they challenge the most basic assumptions, such as what we mean by "soil." The tropical oakleaf basket fern (*Drynaria quercifolia*, not to be confused with the oak fern, *Gymnocarpium dryopteris*) uses sterile leaves to create its own flowerpots, from which the spore-bearing fertile fronds thrust up into the dim light. The "baskets" catch leaf litter and other organic debris, creating an ideal substrate for the fern. Without any contact to the ground, this fern has found a way to thrive in its nutrient-poor environment.

Even after you've worked your way through your identification checklist, some ferns remain a puzzle. In her

Maidenhair fern owes its name to the sensual associations inspired by its delicate, wiry stalks.

Plantlife workshop, Rachel insists this isn't a problem—it's just another reason ferns fascinate us so deeply. The moment you think you've figured them out, they show another side of themselves. Ferns are mischievously varied. In Rachel's fairy-tale forests, the ferns seem darker and more mysterious than the charming decorative plants in Heath's guidebook. Her ferns are shapeshifters, cunning forest sprites, and tricksters, revealing themselves only on their own terms.

Unlike Heath and Rachel Jones, botanists Muriel Bendel and Françoise Alsaker emphasize precise identification skills in their field guide to ferns. In the face of biodiversity loss and rapid habitat change, they argue that accurate knowledge of fern species is crucial. Simply distinguishing ferns from non-ferns or recognizing only the most common fern families is no longer enough. If we want to maintain the current diversity of fern species for decades to come, we need to be able to identify the ferns growing today and take an active role in protecting and managing their habitats. Bendel and Alsaker's field guide adopts the approach of botanical fern societies, which have been meticulously documenting their finds since the nineteenth century. Today, however, such enthusiasm is increasingly tinged by concern—even fear—about the irreversible loss of many species.

This tension between enjoyment and worry, fairy-tale aura and ecological peril, gives the fascination with ferns—long driven by compulsive collecting and aesthetic appreciation—an ethical and perhaps even moral dimension. As in Heath's time, ferns are once again popular in gardens and homes, though not only for human benefit but also to protect the ferns themselves. This makes accurate identification all the more important. The resurgence of fern enthusiasm puts additional pressure on the already endangered habitats of popular species. Knowing if a fern is native or exotic can help people make informed decisions and avoid disturbing habitats unnecessarily. What fern guides and fern identification workshops, as well as historical and current testimonies of fern appreciation, can do is help people see that these elegant fronds are not only beautiful: they are living organisms that need space and deserve our respect.

Fern Secrets

DESPITE THE URGENCY of today's environmental challenges, there is still good reason to be wary of overly rigid classification systems. Outside of the sciences, complex taxonomies don't necessarily deepen our understanding of their subjects. On the other hand, learning to navigate the richness of the natural world directly can help foster a botanical mindset—one that doesn't rely on nomenclature. This mindset is expressed in an inner willingness to let plants interrupt us in our own actions. It becomes outwardly visible when, for example, we suddenly kneel to take a closer look at a fern growing along the path.

Stopping and paying attention are symptoms of a love of ferns that, even without becoming an obsession,

Reduced to their basic shapes, ferns bridge the plant and animal kingdoms. In *Ferns* (2002), artist Roger Matsumoto emphasizes the tentacle-like forms of their root systems and fronds.

can change how we perceive our environment. Once we learn to recognize ferns, we begin to see them everywhere. They leap to the eye, moving from the periphery of our vision to the center and telling their stories. They provide information about their environment and growing conditions and sometimes also about the gardeners, human or animal, who got them there. But unlike trees, with their readily legible bark and rings, ferns are more cryptic: it takes both knowledge and imagination to decipher their stories. Here, botanical and poetic approaches complement each other, creating an intermediate space where we can discover the secrets of ferns without losing our footing in reality.

John Clare's poem "To the Fox Fern" opens up just this kind of common ground. Instead of using scientific knowledge to shed light on the obscure, Clare lingers in the twilight of the poetic. Here fern and poet feel equally at home. But this twilight must not deepen into total darkness. In the imagery of the poem, just enough light or knowledge is needed to perceive the mysterious dimness that surrounds the fern. Clare's poem follows the light, which barely reaches the forest floor. Where it does penetrate as a golden thread, it creates an atmosphere conducive to melancholy reverie—just right for both fern and poet. Too much light—in a lab, a botanist's study, or an overly tidy garden—would destroy the fern, and with it the atmosphere Clare needs for his poetry.

Only in this atmosphere is the fern truly itself: a secretive haunter and trickster, seemingly capable of anything. All of this hinges on Clare's use of the common name "fox fern," now obsolete, and of which no trace remains in the name "hard fern" that has replaced it. The foxtail-like shape of its fronds isn't the only thing the plant shares with its namesake. Like the mythical fox, this plant has as many names as forms, and because it is so easily overlooked, you never know where it might lurk. The romantically evocative mood of the poem grants the fern an agency and independent spirit that elude most attempts at classification. It blurs the categories and boundaries that scientific systems impose on the world, creating a space for the fern as a being whose mysterious otherness exerts an irresistible pull.

This association between ferns and secrets does not originate with Clare or other poets. It long predates aesthetic and botanical interest in ferns. The connection is so close that even the botanists Bendel and Alsaker begin their field guide by noting that ferns evoke "enchanted forests." This is not only because of the dark and humid habitats that many ferns prefer but also because of the peculiar way they propagate. Ferns' biggest secret is how they reproduce. Unlike flowering plants, which display their sexual organs in such a visible and fragrant manner, ferns keep their reproductive mechanisms hidden, subverting modernity's most successful classification system.

This makes fern secrets doubly slippery. It may also help explain why, in the nineteenth century, the seemingly chaste pursuit of ferns was so attractive to young couples.

Already in the eighteenth century, guardians of virtue worried that botanical enthusiasm might lead decent people to indecent thoughts. Carl Linnaeus caused a scandal when he proposed that plants have "male" and "female" reproductive organs. It didn't help that these male and female parts often commingle in the same flower. The ferns' lack of flowers initially seemed to exempt them from such potentially pornographic comparisons. But even Linnaeus suspected that ferns weren't more virtuous than their floral relatives, only more discreet.

Ferns, mosses, and other spore-bearing plants posed a problem for Linnaeus because no one understood how they reproduced. Hence the one "hard name from botany" that Francis George Heath, writing more than a century later, imposed on his readers: Cryptogamia, which literally means hidden (*krypto*) marriages (*gamein*). For pragmatic reasons, Linnaeus placed all plants whose flowers or pollen he could not find into this category of "invisible reproduction." In doing so, he signaled that these plants required further study—a scientific mission that would take decades to complete. Meanwhile, folklore about invisible fern blossoms (magical flowers believed to appear only under special circumstances) persisted, inadvertently reinforced by Linnaeus's classification. The association with mystery has clung to ferns and continues to inspire people interested in magic today, despite the fact that the life cycle of spore-producing plants has long been part of the school curriculum. As intriguing as the fantastical explanations may be, what actually happens beneath a fern's fronds is no less fascinating.

In 1850, the botanist Wilhelm Hofmeister (1824–1877) unlocked the secret of spore-producing plants and was the first to describe the alternation of generations in ferns, mosses, and other cryptogams. Alternation of generations refers to a plant life cycle in which two distinct generations—each with a different form—take turns, one giving rise to the other. While spermatophytes (seed plants) produce seeds that contain all the building blocks for a complete plant, pteridophytes (vascular spore-producing plants) and bryophytes (mosses) produce spores rather than seeds. Spores are haploid: like egg and sperm cells in humans and other animals, they contain only one set of chromosomes. Located on the underside of fronds or on specialized fertile fronds, they mature inside sporangia (spore cases), which eventually burst open, releasing the dustlike spores. These drift through the air and, if lucky, land on a suitable growing surface.

The life cycles of both epiphytic ferns (growing on other plants) and terrestrial ferns (growing in soil) follow the same basic pattern: once the spores land on a suitable substrate, and provided that conditions remain

The prothalli of this fern are heart-shaped—
a fitting form for the "clandestine marriage"
that plays out in the forest shade.

moist, warm, and undisturbed, they begin to develop into gametophytes, plants that don't produce spores themselves but are dedicated exclusively to sexual reproduction. Hardly recognizable as ferns, these tiny prothalli (proto-shoots) consist of a short stem and often heart-shaped leaves. The fact that fern reproduction involves two little green hearts—of all things!—may not be botanically relevant, but the union of form and "content" is too delightful a coincidence to go unmentioned. When the spore germinates, the first phase of the cycle of alternating generations is complete. The second phase is that of the "clandestine marriage" (cryptogamy). The gametophytes produce cells that fuse to form a diploid zygote, from which the sporophyte, or fern, develops. That's how biology textbooks describe it. The fact that these textbooks also use terms such as *egg, sperm, embryo,* and *sexual reproduction* in their descriptions is deliberate, intended to allow a comparison of the reproductive process across species boundaries. But this language is also highly suggestive, not only in a biological sense, and if we had trouble keeping our imaginations in check with the heart-shaped prothalli, things are about to get even more difficult.

In her work on Romantic-era thinking about plants, literary scholar Theresa M. Kelley shows how the concept of "clandestine marriage" inspired Romantic culture and sparked debates in natural philosophy about the "nature of... nature." It offered not only a symbol of clandestine sexual encounters in shadowy woods but also a means to step outside the bounds of order, and not just taxonomically. Kelley suggests that the poet Clare used non-botanical plant names to defy moral norms. The intimacy and directness of his plant writing also characterized his love poems—works often criticized as excessively erotic, their blending of beloved women and plants dismissed as crude. Meanwhile, botanically active women leveraged their scientific study of plants to liberate themselves from what Kelley describes as their ornamental status as "flower women"—delicate and decorative beings needing protection. They, too, operated in the shadowy spaces opened up by the curious status of ferns. Though science had not yet fully unlocked the secrets of ferns, the mere suggestion of a "clandestine marriage" was enough to encourage independent and subversive thinking.

If you follow through on the sexual and erotic allusions, it becomes entirely possible to talk about the alternation of generations in a way that isn't textbook-dry. Things literally become slick when the gametophytes, having developed from spores and now thoroughly wet, seek each other out to fuse together. In an intimate union that generally remains unseen—hidden because of their microscopic size—they combine their genetic material. The metaphorical leap to human encounters is not a large one.

The alternation between asexual and sexual reproduction, the wetness, the darkness, a forest teeming with spores: it all practically cries out to be creatively misunderstood and taken as a cipher for human desire. Ironically, in the 1800s, the Society for Promoting Christian Knowledge welcomed fern enthusiasm and recommended it as a suitable pastime for young women. Relief that ferns kept their sexual organs hidden seems to have made it all too easy to forget that young fern hunters were being sent into dark, damp forests where they might explore more than just botany; after all, chaperones were unlikely to clamber after them to every hard-to-reach spot.

In the second half of the nineteenth century, just as fern fever was peaking, the association of ferns with secret or even forbidden sex emerged in literature as well. It is likely no coincidence that the climactic scene of ill-fated passion in Thomas Hardy's *Far From the Madding Crowd* takes place among ferns. The novel, first serialized in a magazine in 1874, was an instant success and established Hardy's reputation. The story of Gabriel Oak, a suddenly impoverished shepherd, and Bathsheba Everdene, who unexpectedly comes into wealth, is not set in the typical world of fern lovers, but it spoke to them nonetheless. Bathsheba is a striking young woman who singlehandedly runs the farm she has inherited and rebuffs offers of marriage from Gabriel and another, wealthier, suitor. Neither man stirs her passion, and she is determined not to be distracted from her responsibilities.

But then a young sergeant, Frank Troy, turns her head. He's no good for her, of course, and their involvement can only lead to trouble, but he upends her pragmatically ordered existence with a single stroke of his sword. This swordplay takes place against a backdrop of ferns. Frank invites Bathsheba to meet him in a fern clearing at sunset. Bathsheba knows he doesn't want to talk sheep (which she could easily do with Gabriel), and the setting leaves no doubt as to his intentions. The clearing is a hollow covered with "brake fern, plump and diaphanous from recent rapid growth, and radiant in hues of clear and untainted green." The brake fern, or bracken (*Pteridium aquilinum*), tall and lush on this midsummer evening, seems to envelop Bathsheba as she advances into the hollow, its "soft, feathery arms caressing her up to her shoulders." She finds Frank waiting for her, and there, at the center, the outside world disappears behind the "circular horizon of fern." Thus enclosed, the sensible, independent Bathsheba finally surrenders. The fact that Frank wants to display his swordsmanship, and that Bathsheba is left breathless and feeling like "one who has sinned a great sin," suggests that more has happened (or was hoped for) than the kiss that ends the chapter. Soon after, the two are secretly married in the church of a nearby town, officializing the clandestine marriage witnessed by the ferns.

In this study of ferns, shadows dominate, dulling even the vibrant colors of the foxgloves. Ferns thrive in shaded conditions.

Far From the Madding Crowd plays on older fern traditions and combines them with a distinctly critical perspective on the Victorian-era fern craze. The bracken hides the lovers from the rest of the world but doesn't actually make them invisible. Going into the fern hollow, the couple are not seeking botanical forms and variations but a love nest, "floored with a thick flossy carpet of moss and grass intermingled, so yielding that the foot was half-buried within it." This is a hidden allusion to the traditions that around midsummer—the moment of Bathsheba and Frank's meeting—focus attention on ferns.

In the German-speaking world, these traditions center on fern seeds, and in the Baltic countries, on the hunt for fern flowers. In Lithuania, those who find fern flowers become omniscient. But it's no easy quest, since witches, forest spirits, and even the Devil himself want these flowers, too, and actively distract humans from their search. When young Latvian or Estonian couples go out to "find the fern flower," they aren't looking for knowledge but for the same thing Bathsheba and Frank hoped to find under the ferns. Unsurprisingly, "fern flower" is a Latvian euphemism for pregnancy. Today, the Latvian association that provides sexual health services and counseling to young people is called Papardes zieds: fern flower.

With neoromantic wit, Sabine Scho's 2019 poem "farnfrontispiz" (fern frontispiece) captures the secretive nature of ferns and their connection to mystery and poetry, transporting them into the twenty-first century, which is just beginning to discover the cleverness of these plants all over again. Scho's fern seems to mock humans, who pluck spores "from the non-licking side / of its tongue" in order to sell them to the Devil as souls, "somehow glad not to have to / worry any longer about this indefinite part / of their being." In the process, they miss the actual spectacle.

in the other part, the one with the evergreen maidenhair fern
the christmas fern, the male fern, the following drama plays

out: the sporangium tears open and releases the spores
prothalli cling to shady-moist places, male
flagellate cells whip themselves over to the egg cells, all in one
dew-rich night, under tears and fine fluids
the generations alternate and the hart's-tongue fern
says: a seahorse might resemble him more than the
bishop's staff does

To Scho as to Clare, the fern is a subversive actor. In its interactions with and creation of dark, shady, moist environments, the hart's-tongue fern is not only able to speak (speech or language itself being another meaning of *tongue,* as in *mother tongue*); it gives away secrets. These secrets are as provocative as they are subversive, showing that human observers, with their limited perspective, will always struggle to understand the order—or rather disorder—of so-called nature, because they insist on looking for similarity in the wrong places. The hooked shape of a bishop's staff resembles a coiled fern frond, a comparison sometimes used in German for these young shoots. (In English, *crosier*—a synonym for a fern's fiddlehead—also retains the reference.) However, Bischofstab (bishop's staff)—or more commonly Krummstab (crooked staff)—is also the common name of a flowering plant, *Arisarum vulgare*. Being a plant, shouldn't *Arisarum* resemble the hart's-tongue more than a seahorse does? *Arisarum*'s flower consists of a tubular petal that folds over the

The "ancient" ferns in this image by
Jeanne Bieruma Oosting lean toward each other
like trusted friends sharing a secret.

rod-shaped pistil like a cowl. (In English, its common name is friar's cowl.) It looks nothing like a fern. The poem suggests that the difference between animals and plants is no greater than that between flowering and spore-bearing plants. Male seahorses, which also have curled tails, carry their offspring in brood pouches until the young are sufficiently developed to swim on their own. This, too, is a curious reproductive strategy—one that resembles the "drama" of ferns far more than does the very public pollination of angiosperms (flowering plants). Seahorses, like ferns, operate in ways that are secretive and peculiar.

THE NEXT STEPS in the clandestine marriages of ferns are not easy to follow. A few years ago, an interdisciplinary research group finally established how ferns bring their spores into the world. With backlighting and a bit of luck, you can watch ferns release clouds of "dust," but precisely how they do so was long a mystery. While the microscopically small spores travel easily on the wind, their light weight hampers their ability to free themselves from the mother plant (a scientific term that only makes sense if you don't insist on *mother* meaning a female parent). It took the combined effort of physicists and botanists, as well as high-speed cameras and many hours of observation, to arrive at the answer. The sporangia, whether located on the underside of the fronds or on special fertile fronds, act as tiny catapults. As long as they remain moist, everything is stable, but as they dry, these spore cases come under increasing tension. When the spore-bearing fronds of the ostrich fern (*Matteuccia struthiopteris*) change from deep green to an almost sinister black, or, in the study, the pretty round sori of the golden polypody (*Phlebodium aureum*) take on their golden hue, the ring of cells surrounding each sporangium begins to dry out. Because the outer walls of these cells are thinner than the inner ones, considerable tension builds up. As in a catapult, this tension stores energy that will be released in a sudden burst.

When the moisture difference between inside and outside becomes too great, the sporangia split open and expel their spores as far as possible, then immediately snap shut again. The study describes experimental setups with rapid-fire cameras and comparisons with medieval catapults—mechanisms so elaborate that, as a layperson, all I can do is marvel at the design and engineering skills of ferns.

The spores now let themselves be carried by the wind, and if all goes according to plan, they land in "shady-moist places" (Scho). And then? How does the coupling actually proceed? Most articles and descriptions skip over this part, which is probably what makes it so attractive to the poet. It's as if the excessively discreet Victorians still had a hand in the writing, veiling the actual act in silence and allusion, like Hardy did.

But for the sake of a comprehensive fern education, I enlisted two biologist friends of mine, and I now have a fairly precise idea of how this moment in the secret love life of ferns plays out. The prothalli, which are only a few millimeters long, get embedded in moist soil or on slimy films on the ground, branches, or stone walls. The egg cells signal their location by releasing chemicals into this medium, so that the sperm cells, which are equipped with flagella, know where to go. This process, called sperm chemotaxis, was first documented in ferns in 1884. The botanist Wilhelm Pfeffer vividly described how the spermatozoids (Samenfaden, literally "seed threads") could be attracted—"as is well-known, very easily," by placing the prothalli in water—and how he watched them for hours, luring them with malic acid and definitively establishing that the tiny cells could be induced to swim in any direction, provided it smelled right and the medium was suitable. The same, incidentally, is true of the sperm of most mammals. If everything remains shady and moist, the swimming cells stand a chance of reaching their destination. When fertilization occurs, the gametophytes die off and make way for the fern, which emerges from the slime to unfurl its great fronds and start the whole cycle anew.

Fern Symbolism and Pacific Kinship

WHILE FERNS GUARD their secrets closely, shrouding aspects of their basic biology in mystery, our knowledge of their abilities and functions in relation to humans has been obscured through both forgetting and active suppression. This is particularly true of fern-human interactions outside Europe, some of which continue to challenge Western conceptual frameworks to this day. Of the over ten thousand known fern species, only a small number are native to Europe. Most species are indigenous to the tropics, where the year-round humidity and warmth provide ideal growth conditions. Costa Rica, for example, despite its small land area, is home to about 900 species, compared to around 450 in North America and just 170 in Europe. Ferns are used for medicinal purposes almost everywhere they grow, and their fiddleheads—the curled young fronds of certain species—are considered a culinary delicacy in many parts of the world. However, because fiddleheads are difficult to transport and can be highly toxic when spoiled or moldy, these culinary traditions tend to be highly localized. In Canada and the northern U.S., for instance, fresh fiddleheads are usually available only at farmers' markets in spring, near the areas where they grow. An exception is South Korea, where gosari (dried fiddleheads) are a grocery-store staple.

Particularly in the South Pacific, the great density and diversity of ferns have given rise to a fern culture with both practical and symbolic dimensions. Its full scope, however, is only becoming visible again now in the context of decolonization and the reconstruction of Indigenous culture.

No country is more closely associated with ferns than New Zealand. Although Aotearoa, the Māori name for this island nation in the southwest Pacific, spans subtropical and temperate zones, more than two hundred fern species grow here. They include tree ferns such as mamaku,

Silhouetted against the evening sky, a tree and a tree fern display strikingly different forms. The contrast captures why tree ferns seemed so exotic to Europeans.

which can grow up to sixty-five feet tall, and water ferns barely a half inch in size. Between these extremes unfolds a fern culture whose symbols are deeply rooted in Pacific traditions and point to a future beyond colonial structures—one that connects botanical and cultural knowledge. This integration offers a path to coexistence between people and plants without exploitation or oppression.

"Imagine," writes Indigenous journalist and gardener Rob Tipa, "landing on a remote beach on the coast of Te Waipounamu (the South Island) 600, 400 or even 200 years ago. All you have is the clothes you are wearing and a few precious possessions you can carry on your back." The vegetation growing to the water's edge is unfamiliar. The giant trees here exist nowhere else on Earth. Indigenous knowledge about which plants play which roles in human life—and which are best avoided—keeps you alive. This is the kind of knowledge Tipa compiles in his book *Treasures of Tāne*. In Māori mythology, it is Tāne, the god of forests and birds, who enabled life by separating his parents, Rangi and Papa, from their embrace. One version of the story tells that Tāne used his legs to push Rangi up and away from Papa, another that he caused trees to grow in order to hold them apart, which also explains why the famous kauri trees in the northern part of Te Ika-a-Māui, the North Island, grow so tall. Once forced apart, his parents became the sky (Rangi) and the earth (Papa). Between them, light and space finally entered the world, and the trees and birds spread out and flourished. But Tāne's brother Tāwhirimātea, the storm god, was displeased by the change, and helped by his children—the winds, tornadoes, rains, and waves—he destroyed Tāne's trees. Still Tāne continued to live among all these beings, and some stories also credit him with creating humanity.

The separation of sky and earth and the forming of humans from clay are themes common to other creation myths; unlike other gods, however, Tāne never withdraws from his world. The world of Māori and Ngāi Tahu (the Indigenous people of Te Waipounamu) operates by different rules than those imposed by the region's Christian colonizers. Responsibility for a person's actions cannot be shifted onto others—traditions of plant knowledge speak of this as well.

The Ngāi Tahu plant knowledge that Tipa gathers in his book does not distinguish between botanical and cultural understanding. The identification of plants is intimately tied to their use for food, medicine, construction, and clothing. But as Tipa indicates, this orally transmitted knowledge is in danger of being lost. Colonizers dismissed the knowledge of Ngāi Tahu, like that of many Indigenous Peoples, "teaching" them to prioritize European systems over their own mātauranga Māori (traditional knowledge). What sets the book apart from many works by non-Indigenous authors is that it does not promise to restore or improve practices for using plants but rather seeks to help preserve relationships between living beings, the land, and ways of life through this specific form of knowledge. Pānako, for example, a family of nutrient-rich edible ferns, are part of the Māori diet. These ferns are also used to treat skin conditions and alleviate teething pain in babies. Today, all three species of pānako are endangered, and their use is reserved to Ngāi Tahu on the South Island and Māori on the North Island—a right that stems in part from the plants' cultural importance in these contexts beyond their role as a food source. Practitioners skilled in traditional knowledge can use them to treat illnesses, foretell the outcomes of conflict and undertakings, and determine the best time to fell trees. When a waka (canoe) is built, the tree stump is covered with pānako to ease Tāne's loss. Since god and forest are inextricably connected, every "use" of plants and animals requires reparations to ensure life's continuity and honor Tāne's gifts.

New Zealand's three most common tree ferns—ponga, mamaku, and kātote in te reo Māori, the Māori language—are part of almost every aspect of life. The pith of the trunk is edible when cooked for forty-eight hours in an umu (earth oven); it has a turnip-like flavor. The young fronds of tree ferns can be eaten as vegetables. Older fronds, especially those of ponga (*Cyathea dealbata,* or silver fern), can be used to fill mattresses or as bedding

The underside of silver fern fronds reflects light, even in dark forests. As a symbol of Aotearoa, they shine far beyond the island nation.

for animals, like the fronds of European bracken. The trunks are rot resistant and were traditionally used as posts for fences and huts. This quality has also endeared them to modern landscapers, with the result that many stands have been overharvested for use in terraces and low-maintenance garden edging. But ponga can be dangerous too. Slivers were once used as spear tips because they could poison and infect even small wounds, causing prolonged pain. As Tipa reports, this quality now grants these plants a measure of revenge against landscapers, since the splinters pierce even thick gloves.

Ponga, a tree fern that can grow up to thirty-three feet tall, gleams against the lush greenery. The striking thing about it is the silvery-white underside of its fronds. Ponga is endemic to Aotearoa, meaning it is found naturally only here. Yet even within this limited geographic range, populations vary, and not all specimens develop the distinctive silvery undersides of the most famous examples. When they do, the upturned fronds, whether fresh or dry, glow so brightly in the dark forest that they can be used to mark paths through the dense undergrowth. The effect is similar to that of reflective road paint at night. When you're unprepared for it, the luminosity is startling. Unlike the ghostly whiteness of beech fern (*Phegopteris connectilis*) in fall, the silver fern's glow seems intentional, making its symbolic appeal instantly obvious.

The artist Friedensreich Hundertwasser, highly regarded for his environmental activism, spent the last decades of his life in New Zealand. His design for an "indigenous" flag featured the koru, a spiral shape based on an unfurling fern frond.

Many of Europe's botanical gardens cultivate ponga as a symbol of New Zealand. New Zealand soldiers were the first in recent history to adopt it as a symbol, displaying it during the colonial period in the Second Boer War to indicate their identity. But the silver fern is perhaps best known as the emblem of New Zealand's rugby team, the All Blacks, who have worn a silver frond on their jerseys since 1905. The silver fern has also been repeatedly proposed as a symbol for a flag to replace the colonial Union Jack. In this, however, the ponga faces competition from another fern symbol: the koru. Its stylized spiral connects

Aotearoa with the wider Pacific region and is used as well by other Pacific Island nations. Inspired by the furled frond of a tree fern, the koru features prominently in Air New Zealand's logo, rendered in white on black (like the silver fern) on the tail of the airline's planes.

The Austrian artist Friedensreich Hundertwasser lived in New Zealand from the mid-1970s until his death in 2000. He acquired considerable property on the North Island with the aim of returning the land to nature. He also designed an alternative flag for New Zealand—one that expressed his political and ecological beliefs, as well as his fascination with spirals and rejection of the straight line. In his words:

> The flag points in two directions, one against the wind and the enemy, which is sharp like a hatchet, the masculine one; and protected by it is one to leeward, with the wind, a feminine one, which is the opposite, opening and unfolding delicately. The flag is conceived as an indigenous flag and is not intended to replace the official ensign of New Zealand.

Hundertwasser intended his design to complement the existing flag. Tipa takes a similar approach to Indigenous plant knowledge, seeking not to supplant botanical knowledge but to open up additional possibilities for coexistence and cooperation. Both practices are antithetical to the use of koru and silver fern as jealously guarded trademarks—FernMarks!—to "build the New Zealand brand." Instead, they assert freedom and independence from European thought while acknowledging that both frameworks can coexist.

In this sense, these fern symbols can serve as potent political statements against nationalism, even when the line between enthusiastic appropriation and hostile takeover is exceedingly fine. In 2022, in commemoration of the seventieth anniversary of Elizabeth II's accession to the throne, a photo of the queen wearing a silver fern brooch was placed, of all places, next to a mamaku fern in the Temperate House at Kew Gardens. The panel explained that on her first visit to New Zealand in 1953, the Queen was presented with the diamond-studded platinum silver brooch by the mayor of Auckland's wife on behalf of the city's women.

The choice of a Māori symbol to honor Elizabeth II, like the use of the fern on the All Blacks jerseys, is fraught with tensions between national pride and colonial heritage. Both royalty and rugby are remnants of the colonial rule against which New Zealand ferns have emerged as symbols of independence and self-determination. However, ferns represent not only national and brand awareness in New Zealand but also near-forgotten traditions whose actual and symbolic roots reach deep into history and across oceans.

In the 1980s, numerous initiatives were launched to protect and recognize the rights, ways of life, and cultural

New Zealand's temperate forests have
an unusually high diversity of ferns. Tree ferns
are a defining feature of these ecosystems.

heritage of the Indigenous population. Ponga and koru belong to the taonga, treasured possessions of the Māori whose protection is guaranteed by the constitution. Also included in the taonga is the right of Māori to record their stories in writing in their own language. As an important symbol, silver fern links contemporary, diverse New Zealand with its Indigenous and colonial history; it also points to a place that, for many Pacific Islanders, is becoming a refuge from rising sea levels and abandonment by the West.

New Zealand's national museum bears the te reo Māori name Te Papa Tongarewa (container of treasures), highlighting the nation's bicultural identity. It offers resources to help identify and distinguish ferns, giving these plants a significant place in the "canon" of national treasures. In addition to photographs and herbaria, the museum also collects valuable books about ferns, such as the collection that King Tāwhiao presented to Canadian doctor J. T. Rennie in 1888 in thanks for his services, and several "blue books" with early cyanotype photographs of ferns. Te Papa also employs botanists to deepen understanding of ferns. New finds such as *Asplenium lepidotum* and unexpected connections in fern history are among the museum's success stories, arising from collaborations between Indigenous knowledge holders and botanists.

Their research has also yielded a new understanding of the Pacific kinship stories of ferns. One example is the identification of New Zealand fern relatives on Borneo. Such discoveries help to relativize the isolation of the Pacific Islands, at least from a deep-time perspective. At issue are not only traveling fern spores but also the symbols, myths, and stories that circulated through the region. The koru shows up in Māori art as kōwhaiwhai (patterns), for instance in tattoos, where it represents the start of a new family line. It appears throughout much of Polynesia, where it symbolizes new beginnings and continuous movement. The fiddlehead-inspired spiral can be found from New Zealand through to Hawai'i. Almost all Pacific islands have their own fern profile and a shared history of seafaring that likely carried stories, symbols, and spores as part of the cargo.

Hawai'i is home to nearly as many fern species as New Zealand. Here, too, numerous species are endemic, including the red 'ama'u (*Sadleria* spp.). The connection between these two island groups was revisited and brought to popular consciousness through Disney's *Moana* (2016). The film tells the story of its young heroine against the backdrop of the sudden end and slow revival of ocean voyaging among Polynesian peoples. Moana is drawn to the ocean, but as the chief's daughter she is required to remain on the island of her birth. Undeterred, she sets out to find and bring back the magical heart of the goddess Te Fiti. The demigod and trickster Maui stole it to impress humans but, in the process, awakened the lava demon

Te Kā. Moana teams up with Maui to restore balance and avert the environmental catastrophe threatening her island.

Moana's island could be Fiji, Samoa, or Tonga, but the film is deliberately vague about location. It focuses instead on elements of a shared Pacific mythology and culture. At the heart of the plot is a pounamu, a New Zealand greenstone also known as jade (nephrite) that is worn as a pendant, with the koru as one of its traditional forms. The fishhook that is so important in the Disney film is another, and small human figures are common as well. In the film, the spiral embodies Te Fiti's heart, and without it, the goddess, who like Tāne oversees growing life, becomes destructive. And yet in the natural cycle represented by the koru, this destruction has its place too. Lava and earthquakes formed the islands from Hawai'i to Aotearoa, and the fertility of volcanic soil accounts for their dense greenery. Ferns are always the first plants to exploit the fertile potential of cooled lava.

On the Pacific Islands, the regeneration of the volcanically destroyed earth through ferns enacts an ancient story on a small scale. The Hawaiian fern species 'ae (*Polypodium pellucidum*) and 'ama'u (*Sadleria* spp.) are volcano dwellers: 'ae appears as a pioneer settler, while 'ama'u and its red fronds follow in its wake, building up humus in which other plants can later grow. Gardeners warn against planting these two species in gardens, because in "normal"

Eugene von Guérard's *Ferntree Gully in the Dandenong Ranges* (1857) reflects the fascination that "exotic" ferns held for Europeans.

Many similar fern gullies in Australia and New Zealand were destroyed for rail and road construction materials or plundered by European fern enthusiasts.

soil they crowd out most other plants and risk dominating the space for years. Similarly, the U.S. National Park Service cautions against uluhe (*Dicranopteris linearis*), a fern species whose rapid growth can obscure trail markers and even entire trails in the rainforest. Uluhe grows at least as fast as bracken but closer to the ground, sometimes concealing holes and roots—a trap setter inclined to show unfamiliar visitors they don't belong. Uluhe regenerates the forest floor, and like the pioneer species 'ae und 'ama'u, it helps green—or regreen—the world. Especially in the Pacific Islands, which are separated by oceans yet connected, the koru also represents the ability to rapidly take root and flourish in ever-new, ever-changing locations. The fern spiral thus remains a symbol of renewal, connecting humans with the more-than-human world.

Return of the Pteridocene

THE PIONEER FERNS on the Pacific Islands can also be read as indicators of a global future that might once again depend on ferns. Muriel Bendel and Françoise Alsaker allude to this possibility in passing when, to illustrate the distances that spores can travel, they cite the 1883 eruption of Krakatoa. Thanks to newly laid undersea telegraph wires, this cataclysmic occurrence on the volcanic island between Java and Sumatra became the first global media event, fixing the world-destroying power of the Earth's interior in the collective consciousness. But just three years later, eleven fern species were found on the

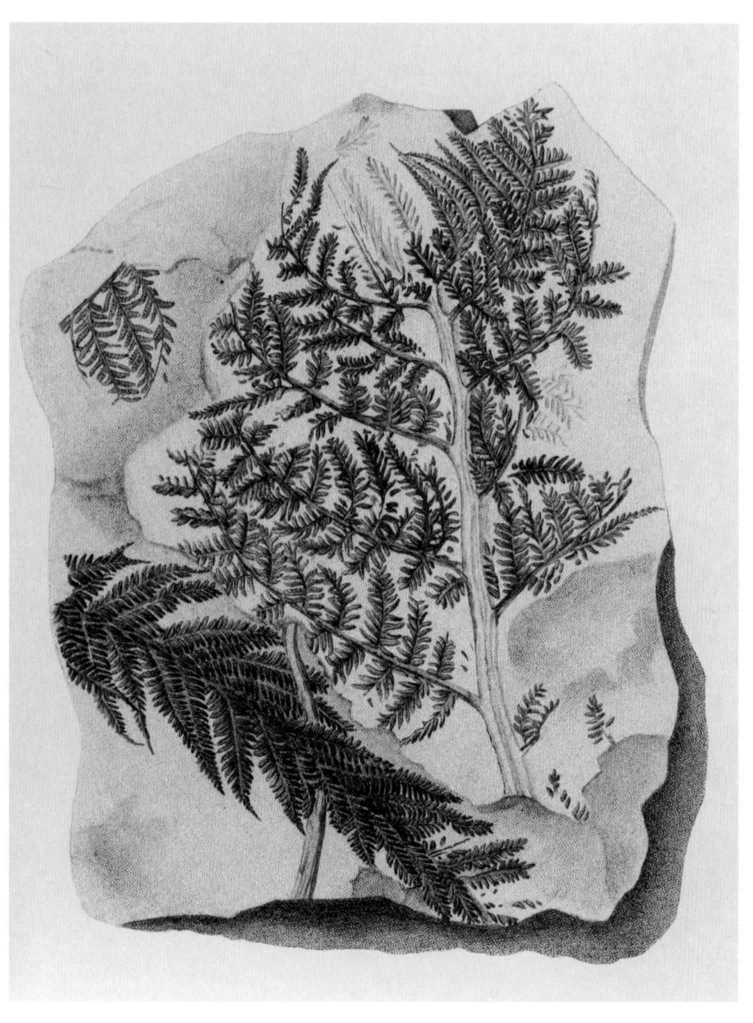

The coal seams that were mined to fuel industrialization yielded an abundance of fern fossils—evidence of the plant origins of this fuel source and a catalyst for broader awareness of Earth's ancient history.

devastated island. Krakatoa is thus also a story about the almost unbelievable resilience of ferns. As in Hawai'i, the ferns proved to be not only capable travelers but also virtually indestructible gardeners of the future, able to transform terrain uninhabitable for humans and animals into land suitable for agriculture, and perhaps even able to found new epochs. There is no reason to assume that we humans will outlive ferns.

The Swedish gardener Anton Sundin's interest in ferns comes from a perspective of ecological sustainability. In his 2019 book about ferns, he begins with a look into the deep past, arriving at the surprising conclusion that "many of the plants we find in nature owe their existence to ferns." Sundin enthusiastically describes how ferns helped usher in a new geological epoch shortly after the dinosaur extinction at the end of the Cretaceous period. Like the Victorians, Sundin is primarily drawn to ferns because of their beauty and the sense of mystery they lend to gardens and interiors. But thinking about deep time has become commonplace since the dinosaur craze that swept North America in the early twentieth century. Combined with today's apocalyptic mood, the scenario he describes conjures an image of a future in which ferns once again assume new relevance.

Although ferns existed long before dinosaurs, from our contemporary human perspective, the fascinating thing about their evolutionary age is that they saw the dinosaurs come and go. In the nineteenth century, the idea that a dominant life-form had ruled the Earth before humans was nearly unimaginable. By the twentieth century, however, it had become a kind of cautionary tale. Unlike the Romans—another historical touchstone for Western cultures—dinosaurs did not build a major civilization and then witness its decline. But the sheer size of the "terrible lizards" (*deinos sauros* in Greek) makes clear that even the most powerful species are not immune to extinction. Alongside that is the overwhelming insight that Earth and life upon it can persist despite cataclysms. Still, underlying any mention of dinosaurs is both our morbid fascination with our own potential demise as a species and awe at Earth's immense age.

But it's not only our preoccupation with extinction that makes the deep-time dimension of fern history so compelling. In *Oaxaca Journal*, Oliver Sacks's account of a fern expedition in southern Mexico, the neurologist and author writes that his love of ferns began at the Natural History Museum in London. Near the dinosaur skeletons, there are fossil gardens displaying fragments of giant club mosses and horsetails that are much older than the museum's star attractions. Dioramas depict what forests might have looked like during that time. Massive fern plants covered continents and reached higher than the longest necks of herbivores. The contrast between these primeval plant giants and their present-day descendants is astonishing. However, Sacks was most impressed by the resilience of ferns.

> Ferns had survived, with little change, for a third of a billion years. Other creatures, like dinosaurs, had come and gone, but ferns, seemingly so frail and vulnerable, had survived all the vicissitudes, all the extinctions the earth had known. My sense of a prehistoric world, of immense spans of time, was first stimulated by ferns and fossil ferns.

There have been five mass extinctions in Earth's history, and all signs suggest we are experiencing the sixth. Ferns have endured at least two of these events, which lends them a certain deep-time gravitas. Garden centers might jokingly label ferns as "dinosaur food" to enhance their marketing appeal, but a historically informed consideration of fern resilience can be genuinely humbling. Maybe that's why Thomas Halliday's 2022 bestseller *Otherlands: A Journey Through Earth's Extinct Worlds* features ferns on its cover. Halliday writes about long-vanished worlds, reconstructing the fauna and flora that made up the ecosystems of various geological time periods. Ferns repeatedly appear in crucial supporting roles, entering the spotlight whenever a world ends.

A team of forty-nine scientists from the U.S., China, Australia, and Germany is investigating the "last frontier

No other plant evokes prehistoric times like ferns do. This has made them a staple of illustrations of primordial worlds, as in this picture of *Rhamphorhynchus* soaring against a backdrop of ferns.

of green plant genomics" to uncover the secret behind ferns' resilience. In a breakthrough in 2022, the scientists deciphered the genome of the water sprite (*Ceratopteris*), finding it to be twice the size of the human genome. However, despite extensive research, we have yet to uncover why this is so, and whether and how ferns make use of this genetic archive. Scientists hypothesize that ferns don't just archive spontaneous mutations but also "take up foreign genes," as stated in a press release from the Justus Liebig University Giessen in Germany. The American project participants were less circumspect, reporting that *Ceratopteris* (nicknamed C-Fern) has repeatedly "stolen" genes for toxins from bacteria and other organisms. C-Fern seems to be something of a kleptomaniac. While this kind of "DNA hoarding" may seem metabolically wasteful in the short term, ferns are perhaps like preppers, stockpiling as a buffer against future disasters.

Especially in the accounts of Sacks and Halliday, ferns appear to sit patiently in the background, letting others rule the Earth until their turn comes around again. Science seems to confirm this. In art and culture, too, ferns assert themselves from the shadows. In many dioramas and pictorial representations, they represent not only the "primordial world" but also its continuity with our own world. Surrounded by ferns, it's easy to dream yourself back into prehistory, carried by this tangible connection to the distant past. "My sense of a prehistoric world, of immense spans of time, was first stimulated by ferns and fossil ferns," writes Sacks, bringing ferns out of the shadows of the dinosaurs.

Yet it is precisely there, in the shadows, that ferns expend their greatest energy. Remnants of Earth's distant past, today's ferns grow mostly unnoticed in dimly lit gardens and woods. Meanwhile, fossilized ferns lie in the darkness underground, yielding energy that we use in the form of coal, petroleum, and natural gas. Not everyone is aware that these raw materials are formed from living organisms—including vast quantities of prehistoric ferns. This ignorance diminishes neither their energy value nor the catastrophic harm, both present and future, that we cause by burning them.

In the Dresden University of Technology's botanical garden, a pond filled with water lilies and surrounded by royal ferns (*Osmunda regalis*), male ferns (*Dryopteris filix-mas*), and other fern species evokes a sense of the primeval. This arrangement grows in the shadow of several coast redwoods (*Sequoia sempervirens*). It will take a few more centuries for the redwoods to reach their full height, but a nearby fossilized tree trunk creates the kind of conceptual bridge that Oliver Sacks describes. Even this subtle nudge is enough to tap into the reservoir of ideas we have about prehistoric times and the vegetation not only beneath our feet, but also heating our homes, albeit via several power stations. In the garden of the research university, these ancient plants stand out as living counterparts to the fossil fuels we can't seem to give up.

This kind of primeval garden is doubly fitting here in the German state of Saxony. Although the region is criticized for mining lignite, a relatively young, low-grade coal, Saxony has strong links to ancient plant life through both its abundance of ferns and its fern fossil treasures. The Sterzeleanum, a collection of fossilized plants from the petrified forest beneath the city of Chemnitz that serves as the main attraction of the city's Museum of Natural History, is well known to locals. But visitors wouldn't necessarily guess that the cultural center that today houses the museum—along with the city library and other services—contains hundred-million-year-old ferns. Almost three stories high, the horsetail tree (genus *Calamites*) stems in the atrium of the Tietz building were previously displayed in front of a museum built specifically for the paleontological collection. Now they rise like giant cathedral pillars in an otherwise mundane structure.

Oliver Sacks's astonishment at the extreme difference in size between prehistoric horsetails and contemporary calamites, which grow at most to hip height, becomes instantly comprehensible when you see these giants. Scientists haven't definitively identified them as calamites, but they have determined that they are spore plants. The contrast in scale is staggering: imagining the crowns and roots of these colossal horsetails is as difficult as picturing chicken- and lizard-like creatures the size of semitrailers.

The museum, which is devoted primarily to fossilized plants, helps visitors grasp the transformations that have occurred over time. The display also includes historically significant finds that document fossil-collecting practices dating as far back as the eighteenth century. Many

Before being relocated to Chemnitz's cultural center, the petrified trees stood in front of the city's former natural history museum as evidence of the Zeisigwald volcano, which buried the surrounding forest 291 million years ago.

are petrified wood—ancient trees that mineralized and turned to stone over time. Chemnitz's petrified forest has attracted attention since at least the Middle Ages. Georg Bauer, better known as Georgius Agricola, served as town physician and later mayor in the mid-1500s. He also coined the term *fossil* in his ten-volume opus *De natura fossilium* (roughly translated as "On the nature of things dug out of the earth"), published in 1546—a work that confirmed the stones' worth as objects of scientific interest. During Agricola's time, *fossil* referred to anything found in the earth, including living organisms that had become fossilized and the impressions they left behind, as well as inorganic minerals. However, with the discovery of Earth's immense geological age in the eighteenth century and the development of evolutionary concepts from Lamarck to Darwin, the term *fossil* narrowed in meaning to signify only traces of life from the geological past.

Researchers continue to study the evolutionary history of this area today. Recently, a 290-million-year-old seed fern fossil was unearthed in the city of Chemnitz. Referred to in newspapers as a "giant fern," the plant represents an extinct intermediate link between ferns and seed plants. Dating to the Permian period, *Medullosa stellata* was widely distributed in the Chemnitz forest, yet it remains an enigma. It embodies one of nature's most impressive experiments, because it has both fern fronds and seeds—features that seem contradictory from our perspective. While other places saw the spread of gymnosperms—trees whose seeds are not enclosed within fruits, a category that includes all conifers—nature tried something entirely different here. Though the seed fern model did not prevail, its existence is notable (and delightfully tantalizing, given the many myths and traditions about invisible fern seeds). As it turns out, the problem is not that fern seeds are invisible, but that they died out.

Unlike fern seeds, fern fossils are easy to find. As a child, Oliver Sacks learned from his mother that the coal heating their home was largely composed of ancient ferns—and that coal deposits were a rich source of their fossilized remains. Coal was generally considered more valuable as fuel and raw material than as a source of museum specimens, but the industrial demand for coal drove extensive mining operations that then made fern fossils widely accessible. Even in Germany, which was relatively untouched by fern fever, these abundant relics stood as incontrovertible evidence of an unimaginably distant past.

Fossil ferns symbolized the triumph of the present over the past on a planetary scale. In other words, they represented the ability of humans to harvest prehistoric plants and use them to fuel progress, sparing themselves from the fate that overtook dinosaurs. Every image showing a dinosaur surrounded by ferns thus also conveyed that humans were able to turn even the greatest tragedy to their advantage and reinforced their genuine belief that they could create their own climate. The fact that this climate would likely resemble a new "Pteridocene" (fern age) rather than an Anthropocene demonstrates that even deep time is fair game for historical irony.

The dappled light favored by ferns gives woods a fairy-tale quality. Where forest management is guided by more than economic criteria, ferns flourish alongside trees, mosses, and non-vegetal beings to create vibrant, biodiverse communities.

One of my fern rambles took me to Saxon Switzerland, a mountainous region south of Dresden, where I visited the famous gorge known as the Uttewalder Grund. The ravine's cellar-like darkness and humidity suit ferns perfectly. The walls, several stories high, bristle with ferns, and the light is almost as green as it is in Devon. Yet these rugged cliffs tell a different story than the hills and rainforests of southern England, though they're no less romantic. The days when German landscape painter Caspar David Friedrich found this place shrouded in fog and darkness are long past. The Teufelsküche (devil's kitchen) and various other caves are still charming, but one thing has become clear in recent summers: even here, moist refuges are becoming scarce.

Today, the ferns grow in a parched streambed, their fronds rustling because they, too, are parched, while the moss glistens only in the few spots where mountain springs still seep out of the rocks.

I was there on a weekend before the big wildfires that marked the summer of 2022 in this region near the German-Czech border. Rising temperatures have already pushed many fern species in the area to the brink of extinction, and too many no longer grow here at all. I didn't bother looking for hart's-tongue, for instance—decades of increasing dryness have eliminated it from this landscape. Ferns face similar pressures elsewhere. In the Pacific Northwest near Seattle, the once-abundant sword ferns are disappearing from forests. In the heart of Europe, the range of most fern species is contracting. Threats to forests also affect ferns. However, disruptions also cause openings. In areas clear-cut after bark beetle invasions, bracken is often the only plant that can still grow. Without human intervention, it takes decades for other plants to gain a foothold.

In recent years, the tree canopy in the Saxon Switzerland National Park has noticeably thinned. Yet at the heart of the park, the fern-world feeling remains. The ferns are so present—so distinctively green, lush, and unconcerned with how we expect plants to behave—that everything seems entirely normal. They move quickly, grow everywhere, and don't seem to care whether they "belong." It's another reason to pay attention to ferns. They remind us that when we protect forests, we protect not only trees but also entire worlds that could not otherwise exist. Ultimately, I'm not too worried about ferns. Their history suggests that they will always be the first to grow back. I just hope we get to be part of that story.

Portraits

Male Fern

Dryopteris filix-mas

WHEN YOU PICTURE a fern, chances are you're thinking of male ferns. With their quintessentially fern-shaped fronds, they grow almost anywhere sufficiently damp. Orderly rosettes produce upright fronds ranging from two to five feet tall, forming large funnels that sometimes remain green well into winter. Their leaf blades are deeply lobed, with leaflets that start small at the base, widen gracefully toward the middle, and taper to the characteristic pointed fern tip. Male ferns thrive in cool wooded areas and on the banks of streams and rivers, and they can colonize railway embankments if conditions are moist enough. The sori on the undersides of the leaflets cluster near the narrow stems and are covered by kidney-shaped indusia in early summer.

The wormlike appearance of these plump spore cases may have inspired the traditional use of *Dryopteris* to expel tapeworms in humans and protect stored apples

from pests. When the sori are ripe, the indusia shrivel, releasing the spores, which path-side ferns disperse onto anything that brushes against them. *Dryopteris filix-mas* was long considered the botanical counterpart to the lady fern (*Athyrium filix-femina*), hence the name male fern. An association with romance still clings to these plants, as in Latvia, where young couples hunt for "fern flowers" as part of traditional midsummer celebrations.

Hard Fern, Deer Fern

Blechnum spicant

THESE EVERGREEN WOODLAND PLANTS tend to be solitary, with their clumps of long narrow fronds often spaced far apart. This spacing makes them easy to spot, even without the pronounced funnel shape that characterizes male and ostrich ferns. Hard ferns are dimorphic, producing two distinct types of fronds: spore-bearing fertile fronds and photosynthesizing sterile ones. The sterile fronds are pinnatifid, meaning their leaflets are not further divided. Leaflets of similar length alternate on either side of the stem in a slightly offset pattern. Bright green when young, these fronds become leathery and dark by summer's end. In winter, they almost resemble spruce needles, blending perfectly with their wooded surroundings. The longer spore-carrying fronds rise upright from the center of the rosette and are a lighter shade of green. The Swiss name Leiterlifarn (ladder fern) aptly describes their shape. An

older English common name, fox fern, likely refers either to their wooded habitat or their foxtail-like fronds.

Maidenhair Fern, Venus Hair Fern

Adiantum capillus-veneris

MAIDENHAIR FERNS MAKE UP a large family of particularly delicate ferns widely distributed throughout tropical regions. Instead of the sword-shaped fronds typical of many ferns, their leaf structure divides into fan-shaped leaflets that hang from delicate hairlike stems. *Adiantum capillus-veneris,* the only maidenhair native to Europe, also grows in North America, primarily in the southern U.S. Found in caves and on wet cliffs and canyon walls, it can cover entire rock faces to enchanting effect.

Adiantum means "unwetted"—a reference to the plant's water-resistant leaves—but it is the hairlike stems that seem to have inspired its name. The allusion to Venus's hair (*capillus veneris*) likely links the plant's love of moist habitats with other, normally hidden, wiry dark hair, another association of ferns with lust. This

gives maidenhair ferns, with their preference for out-of-the-way spots, a charmingly suggestive quality that has not harmed their appeal. During the Victorian fern craze, it made them sought-after collector's items and fashionable decorative motifs. Today, maidenhair ferns are popular once again as graceful, albeit temperamental, houseplants prized for their cascading foliage.

Hart's-Tongue Fern

Asplenium scolopendrium

HART'S-TONGUE FERNS ARE easily recognized by their undivided fronds. These mutation-prone plants are variable in form, with frond edges ranging from smooth to baroquely ruffled, but all have in common an absence of leaflets. Their strap-like shape resembles that of tropical species like the bird's-nest fern (*Asplenium nidus*), an epiphyte that favors tall trees, and crocodile fern (*Microsorum musifolium* "Crocodyllus"), today a common houseplant. Hart's-tongue ferns thrive in calcareous soil, for instance in coniferous forests, and serve as an indicator species for increasingly rare ravine woodlands consisting of alder, maple, and linden. They have an affinity for rock ledges and caves, and especially stone walls and hedges where limestone is present. Depending on their environment, they form fronds ranging in length from two to twenty-three inches.

Traditionally used to treat digestive problems, they feature in the medical writings of Hildegard of Bingen. The medieval naturalist recommended taking hart's-tongue after breakfast, either brewed with wine, milk, and spices or in powdered form, licked from the hand.

Oakleaf Basket Fern

Drynaria quercifolia

THIS EPIPHYTE HAS solved an ecological problem with extraordinary but underappreciated design ingenuity. In the tropical rainforests of India, Southeast Asia, and especially the Philippines, where it is known as kabkab, this species would starve without its own flowerpot system. Kabkab grows mostly on trees and has dimorphic fronds, meaning it produces two distinct frond types. The fertile fronds grow on long stems and stretch their simply divided leaf blades toward the light, with sori clustering in small spots on their underside. These fronds therefore fulfill two functions: they capture light and ensure reproduction. The sterile fronds, meanwhile, cling to the tree trunk like small cups. When their leathery leaves dry up, they resemble European oak leaves in fall—which explains the plant's epithet, *quercifolia* (from the Latin *quercus* for oak, and *folia* for leaf). With a texture between

leather and paper, these "cups" catch everything that falls from above, holding the debris to create just enough humus to nourish kabkab's long green fronds as they reach far into the forest.

Tree Ferns

Cyatheales (order)

IN THE CASE of ferns, as with trees generally, the term *tree* describes a growth form rather than a species. Across the southern hemisphere, especially in tropical regions, some funnel-shaped ferns develop trunks that lift their crown of fronds high above the ground. Their relatively slender trunks distinguish them from "true" trees like oaks, maples, and firs. These trunks form as the plant sheds fronds from previous growing seasons. Fossil tree ferns have been found worldwide. At various times in the Earth's history, they grew alongside treelike horsetails, forming towering forests that are now mainly remembered as dinosaur food. Modern tree fern species include the Australian pocket ferns (genus *Dicksonia*) and New Zealand's scaly ferns (*Cyathea*).

Two especially impressive species are mamaku (*Cyathea medullaris*) and ponga (*Cyathea dealbata*). Mamaku can exceed sixty-five feet, making it the tallest of today's

tree ferns, and was long prized for its edible pith. Ponga, or silver fern, is easily recognizable by the silvery-white underside of its fronds. Bright enough to serve as night-time path markers in the bush, the fronds have become New Zealand's national emblem. Ponga's fibrous trunk, once used to make spear tips, now provides material for rot-resistant decking. Both mamaku and ponga ferns develop crowns with fronds up to sixteen feet long, easily rivaling those of large palm trees.

Water Fern

Azolla filiculoides

INDIVIDUAL WATER FERNS measure just a fraction of an inch, but they have collectivist tendencies, clumping together to form vast mats that can extend over many acres. Their extremely rapid reproduction rate, combined with their ability to hitch rides on waterbirds across long distances, makes them a formidable threat. In warm and temperate regions in the Americas, water ferns are feared as invasive species capable of blanketing lakes and suffocating all life below, devastating aquatic ecosystems and leaving behind only a foul-smelling broth. In the fall, the large floating mats of *Azolla* turn a notorious red. *Azolla* has also emerged as an invasive concern in Central Europe, where Western Germany has already classified it as a problem plant.

But that's just one side of the story, because *Azolla* is also considered an environmental savior. Water ferns live in symbiosis with cyanobacteria that fix nitrogen

from the air. This forms tiny air pockets under the leaves, making water ferns both good swimmers and an excellent green fertilizer. In Asia especially, *Azolla* is valued as a sustainable alternative to synthetic fertilizers. During and after the rice harvest, farmers deliberately allow the water fern to spread. When the ferns eventually die back, they replenish the depleted soil with nitrogen.

Bracken, Brake, Eagle Fern

Pteridium aquilinum

BRACKEN IS EASILY recognized by its tall upright fronds, which emerge along underground rhizomes. In woodlands, the plants often look as if they're standing in neat, orderly formation, their fronds unfurling from clearly branched straight stems. The name *Pteridium* (from the Greek *pteron,* for feather) refers to the appearance of bracken's fronds, which unfurl in late spring as variously sized fiddleheads at the ends of these branched rachises. Bracken also stands out in fall, creating vibrant displays ranging from pale yellow to rust red.

Bracken typically forms dense colonies that dominate the areas they occupy. After clear-cutting and forest fires, on abandoned construction sites, and along roads and highways, bracken spreads rapidly through its highly productive rhizomes, suppressing other growth by literally

overshadowing it. This makes it both a competitor to forestry and a threat to grazing lands, despite its usefulness for erosion control and as livestock bedding. Problems arise when warming temperatures remove the constraint on its frost-sensitive rhizomes, allowing it to spread unchecked. Thus freed, bracken capitalizes ruthlessly on the environmental disruption that humans create.

References

Bächtold-Stäubli, Hanns, and Eduard Hoffmann-Krayer, eds. "Farn." In *Handwörterbuch des deutschen Aberglaubens*. Walter de Gruyter, 1987.

Bate, Jonathan. *John Clare: A Biography*. Farrar, Straus, and Giroux, 2003.

Bendel, Muriel, and Françoise D. Alsaker. *Farne, Schachtelhalme, und Bärlappe: Der Naturführer zu den Farnpflanzen Mitteleuropas*. Haupt Verlag, 2021.

Clare, John. *John Clare: Selected Poems*. Edited by Jonathan Bate. Faber & Faber, 2004.

Clare, John. *A Language That Is Ever Green*. Edited by Manfred Pfister. Verlag Das Kulturelle Gedächtnis, 2021.

Halliday, Thomas. *Otherlands: A Journey Through Earth's Extinct Worlds*. Random House, 2022.

Hardy, Thomas. *Far From the Madding Crowd*. Penguin Books, 2003. Originally published in 1874.

Heath, Francis George. *The Fern Paradise: A Plea for the Culture of Ferns.* Hodder and Stoughton, 1875.

Kelley, Theresa. *Clandestine Marriage: Botany and Romantic Culture.* Johns Hopkins University Press, 2012.

Keogh, Luke. *The Wardian Case: How a Simple Box Moved Plants and Changed the World.* University of Chicago Press, 2020.

Mabey, Richard. *Flora Britannica.* Sinclair-Stevenson, 1996.

Sachsse, Rolf. *Anna Atkins: Blue Prints.* Hirmer Verlag, 2021.

Sacks, Oliver W. *Oaxaca Journal.* Vintage Canada, 2012.

Steffen, Richie, and Sue Olsen. *The Plant Lover's Guide to Ferns.* Timber Press, 2015.

Sundin, Anton. *Farne: Vielfalt und Geschichte einer der ältesten Pflanzengruppen.* Haupt Verlag, 2023.

Teltscher, Kate. *Palace of Palms: Tropical Dreams and the Making of Kew.* Picador, 2020.

Tipa, Rob. *Treasures of Tāne: Plants of Ngāi Tahu.* HUIA Publishers, 2018.

Weinstein, Mobee. *The Complete Book of Ferns: Indoors, Outdoors, Growing, Crafting, History & Lore.* Cool Springs Press, 2020.

Whittingham, Sarah. *Fern Fever: The Story of Pteridomania.* Frances Lincoln Adult, 2012.

Whittingham, Sarah. *The Victorian Fern Craze.* Shire Publications, 2009.

List of Illustrations

6 C. E. Faxon and J. H. Emerton, "Ostrich-Fern," reprinted from *Beautiful Ferns: From Original Water-Color Drawings After Nature* (D. Lothrop and Company, 1882), 47.

9 Karl Blossfeldt, "*Aspidium filix mas.* Wurmfarn. Junge gerollte Blätter in 4facher Vergrößerung," reprinted from *Urformen der Kunst* (Wasmuth Verlag, 1928), 56.

14 W. Müller and C. F. Schmidt, "*Aspidium filix mas* Sw.," reprinted from *Köhler's Medizinal-Pflanzen in naturgetreuen Abbildungen*, vol. 1, ed. Gustav Pabst (Gera-Untermhaus, 1887–98), plate 81.

19 John White Abbott, *Canonteign, Devon,* 1804, watercolor with pen and gray ink on paper, 23.7 × 29.4 inches (60.3 × 74.6 cm), Yale Center for British Art, Paul Mellon Fund.

53 "Centre of the Great Palm House at the Royal Botanic Gardens of Kew," reprinted from *The Illustrated London News*, August 7, 1852.

58 Helen Paterson Allingham, "Gathering Ferns," reprinted from *The Illustrated London News*, July 1, 1871.

61 Reprinted from Francis George Heath, *The Fern Paradise: A Plea for the Culture of Ferns*, 4th ed. (Sampson Low, Marston, Searle, and Rivington, 1878), plate 1.

68 "British Ferns," reprinted from *The Boy's Own Paper*, ca. 1885.

73 Anne Pratt, "Rigid Three-Branched Polypody, *Polypodium calcareum*," reprinted from *The Flowering Plants, Grasses, Sedges, & Ferns of Great Britain and Their Allies the Club Mosses, Horsetails, &c*, vol. 4 (Frederick Warne & Co., 1905), plate 282.

78 Henry Bradbury, "*Adiantum capillus-veneris*," reprinted from *The Ferns of Great Britain and Ireland*, by Thomas Moore (Bradbury and Evans, 1855), plate 45.

82 Roger Matsumoto, *Ferns*, 2002, palladium print, 20.5 × 14.8 inches (52.1 × 37.6 cm).

87 Arnold Dodel-Port and Carolina Dodel-Port, "*Aspidium—Prothallium*," reprinted from *Anatomisch-Physiologischer Atlas der Botanik* (J. F. Schreiber, 1878).

92 Hans Gude, *Study of Ferns*, 1846, oil on cardboard, 13.2 × 10.6 inches (33.5 × 27 cm).

126 Friedrich Gahlbeck, *Chemnitz, Versteinerte Bäume*, 1964, photograph, © Bundesarchiv.

129 Ivan Shishkin, *Stones in the Forest, Valaam*, 1858–60, oil on canvas, 12.4 × 16.9 inches (31.7 × 43 cm).

134–49 Maps and illustrations © by Falk Nordmann, 2024.

Index

Illustrations indicated by page numbers in italics

A

B

C

D

E

F

G

H

U

V

W